SUPER HUMAN

SUPER HUMAN

THE BULLETPROOF PLAN TO AGE BACKWARD
AND MAYBE EVEN LIVE FOREVER

DAVE ASPREY

HARPER WAVE

An Imprint of HarperCollins*Publishers*

A portion of the proceeds from the sale of this book have been donated to the XPRIZE Foundation to support the creation of a better world in the future . . . because we're going to be here to enjoy it.

HarperCollins books may be purchased for educational, business, or sales promotional use. For information, please email the Special Markets Department at SPsales@harpercollins .com.

FIRST EDITION

Library of Congress Cataloging-in-Publication Data has been applied for.

ISBN 978-0-06-288282-0

19 20 21 22 23 LSC 10 9 8 7 6 5 4 3 2 1

To my children, Anna (twelve) and Alan (nine), who diligently sat by my side and edited this book in a way that genuinely helped make it better. It's my sincere hope that when you are both over a hundred years old, you'll let me help you edit whatever you're creating. I plan to be there for you.

Do not go gentle into that good night,
Old age should burn and rave at close of day;
Rage, rage against the dying of the light.

—*Dylan Thomas*

CONTENTS

INTRODUCTION: YOUR ANCESTORS WERE BIOHACKERS XIII

PART I: DON'T DIE

1 THE FOUR KILLERS 3

2 THE SEVEN PILLARS OF AGING 24

3 FOOD IS AN ANTI-AGING DRUG 41

4 SLEEP OR DIE 63

5 USING LIGHT TO GAIN SUPERPOWERS 84

PART II: AGE BACKWARD

6 TURN YOUR BRAIN BACK ON 103

7 METAL BASHING 129

8 POLLUTING YOUR BODY WITH OZONE 144

9 FERTILITY = LONGEVITY 156

10 YOUR TEETH ARE A WINDOW TO THE NERVOUS SYSTEM 174

11 HUMANS ARE WALKING PETRI DISHES 185

PART III: HEAL LIKE A DEITY

12 VIRGIN CELLS AND VAMPIRE BLOOD 209

13 DON'T LOOK LIKE AN ALIEN: AVOIDING BALDNESS, GRAYS, AND WRINKLES 232

14 HACK YOUR LONGEVITY LIKE A RUSSIAN 248

AFTERWORD.. 265

ACKNOWLEDGMENTS 267

NOTES 269

INDEX 311

INTRODUCTION:
YOUR ANCESTORS WERE
BIOHACKERS

A hundred thousand years ago, two cavemen struggled to keep their families alive during a particularly harsh winter. As the wind howled, one wrapped himself in animal skins, checked that the fire was big enough to keep his family from freezing, and made the dangerous trek to a neighboring cave. He ducked his head to avoid banging his overhanging brow at the entrance, shivered as he noticed the dark cave was scarcely warmer than the air outside, and shouted excitedly, "Thog, I have discovered something amazing. You have to see this!" Thog reluctantly wrapped himself in animal skins and ventured into his neighbor's impossibly warm and well-lit cave, where he saw the world's very first man-made fire. "Isn't this incredible?" the caveman said. "I am using this *right now* to keep my cave warm. See how happy my kids are? Do you want me to show you how I am doing it?"

Thog was skeptical. He *knew* fire was dangerous. When lightning struck a tree, the resulting wildfire could burn forests, not to mention humans who were dumb enough to get too close. He and all the other cave dwellers had survived winter (for the most part) without fire. They huddled together and shared their food, and everyone got along. Fire might be harder to share. What if only some cavemen had access to its warmth? "No thanks," Thog grunted. "I'm good." And he shivered his way back to his cold, dark cave.

One of those guys is our ancestor. And—spoiler alert—it's not Thog.

Fire was one of the first tools humans discovered to help extend our life-spans, and we've been searching for new and increasingly complicated tools ever since. We have a hardwired instinct to avoid death that predates written language and even our ability to stand upright. Our awareness of our own mortality has led us to innovate throughout millennia to avoid dying, which of course means living longer. It is the fundamental drive of the human race, it is what has allowed us to evolve as a species, and we are nowhere near done.

Fast-forward from our caveman ancestor to the beginnings of recorded history, and we find proof that humans have been seeking immortality since we started writing things down. About 2,400 years ago, the pharaohs of Egypt in Alexandria devoted an enormous amount of their wealth and power to a quest for "eternal life." In China, Taoist philosophers placed a tremendous amount of value on longevity. To achieve it, they focused on internal alchemy (visualizations, nutrition, meditation, self-control, and even sexual exercises) and external alchemy (breathing techniques, physical exercises, yoga, medical skills, and producing an "elixir of immortality" using various purified metals and complex compounds). In India, the theme of prolonged life emerged in Ayurvedic texts as rasayana, the science of lengthening life-span.

You could say to yourself, "Great, a couple thousand years ago, some crazy people wanted to live a long time. They're dead now." Except . . . these life-extending self-proclaimed alchemists are part of a lineage of biohackers that includes some of the most influential forefathers of modern science and medicine, such as Isaac Newton, Francis Bacon, Paracelsus, Tycho Brahe, and Robert Boyle. (Unfortunately, most female alchemists are not well known because they were accused of practicing witchcraft and killed.) The quest to live longer drove the scientific revolution, and it's reasonable to say that the technology you rely on today would not exist without our core drive to live longer.

Along the way, charlatans and con artists took advantage of the burgeoning market of life-span extension by selling people on the idea of turning lead into gold. Soon alchemy itself was redefined as "false magic." Today it conjures images of wizards in pointy hats. But the reality is that early alchemists were seeking something most of us would gladly trade our gold for: immortality. Humans have literally been working on transmuting our species from mortal to immortal for thousands of years. I'm one of them, and this book is about what it's been like to work on extending my own life for the past twenty years.

The game has changed now that we have access to more knowledge and data than ever before. Not dying is still the number one motivator for all humans, and it isn't because we choose it. This desire is baked into us at the subcellular level to the point that avoiding death is automatic. As I was researching my last science book, *Head Strong*, it became clear that our innate drive to avoid death comes from deeper within us than you might expect.

Your mitochondria, the power plants in your cells that evolved from ancient bacteria, have the same basic goal of any successful life-form—to stay alive. The human body has at least a *quadrillion* mitochondria scurrying around inside it, each one of them running a program that says, "Don't die." Is it any wonder, then, that you don't want to die? Those ancient bacteria drive you to focus on behaviors that will keep your meat alive and able to reproduce. I call these behaviors the three Fs: *fear* (fight off or flee from things that might kill you), *feed* (eat everything in sight so you have energy to fight off or flee from things that might kill you), and the other f-word that propagates the species. You spend a lot of time on these three priorities, don't you?

All life-forms—from bacteria to fruit flies to tigers—share the same basic instincts, but we're the only ones with big enough brains to also make long-term decisions to support our goal of not dying. Ironically, we are often distracted from making good long-term decisions for our longevity by the very instincts that are meant to keep us alive. For example, our desire not to die from starvation leads us to consume too much sugar for a quick boost of energy. This keeps us alive in the short term and increases our chance of dying in the long term. To have a perfectly functioning body and mind long past the age when you can no longer reproduce (at which point you essentially become useless to your mitochondria), you must build practices that prevent you from falling prey to those base instincts that make you a short-term thinker.

So if we've been seeking immortality for centuries and this drive comes from deep within our biology, why do people laugh when they hear I'm planning to live to at least a hundred and eighty? Some people stop laughing when they see I'm dead serious (no pun intended), but many act like Thog, shivering their way back to their dank caves.

We've already seen that it's possible to live to a hundred and twenty. The longest verified living person made it to a hundred and twenty-two, and there are scattered but unverified reports of a hundred and forty. Over the last twenty years, the rules in the anti-aging field have clearly changed. If you make good daily decisions that benefit longevity and pair those choices with new technologies that can prevent and reverse disease and aging, it is becoming possible to add at least 50 percent to the age of the longest-lived human. Hence, living to a hundred and eighty is a realistic and achievable goal, at least if you're willing to do the work along the way to get there. The good news is that even if I'm wrong, I'll get to enjoy however many years I do have a whole lot more thanks to these practices. If in the end they only help me avoid Alzheimer's or buy me an extra year with the people I care about, it's still a win in my book.

These daily decisions and interventions are investments in my future, but they also power my performance right now. Each has its own return on investment (ROI). Some, like eating the right foods and getting quality sleep, may provide a longevity return of 3 percent, along with a better brain right now. Others, like fixing my jaw alignment or strategically using lasers on my brain, might yield closer to a 6 percent return on longevity. Some of the most radical, such as consuming oil containing an unusually shaped carbon molecule that helped rats in a lab live 90 percent longer than expected, may have incredibly high returns . . . if they work at all and don't cause unintended harm if they fail. Today it is difficult to calculate exactly what longevity return you might receive on a specific intervention, but we do know the ROI comes in the form of more energy *now* and years of better health *later*. These are not just any years, but quality years filled with energy and mobility and brainpower, plus the wisdom that comes with living well for so long.

This type of energetic, productive old age is difficult to imagine, which is why many people shudder at the thought of living to a hundred and eighty. When I interviewed Maria Shriver on my podcast, *Bulletproof Radio,* her response to my mission was "I don't want to live to a hundred and eighty. You can have that!" Most of us so badly want to avoid the picture we have of old age—suffering from chronic pain, becoming house- or wheelchair-bound, helplessly relying on care from others, forgetting our loved ones' names—that we would rather die. Me too. But it doesn't have to be that way, and I've been blessed to interview and befriend a great number of Super Humans who are not only thriving, but also happily giving back to society in their seventies, eighties, and even nineties.

See, not dying will only get you so far. That's step one. But living *longer* doesn't necessarily mean living *better.* Step two is gaining the energy you need to stop aging in its tracks and start aging backward. Step three is the icing on the cake that takes you from mere mortal to Super Human: someone with the wisdom of age but who heals and regenerates like a teenager. This, too, has been a human goal throughout history. Look no further than the Fountain of Youth, which first appeared in writings by Herodotus, an ancient Greek historian, in the fifth century BC. Herodotus claimed there was a fountain with magical, longevity-promoting water in the land of the Macrobians, a legendary race of people who all lived to be . . . a hundred and twenty. There's that number again.

Interestingly, in his writings, Herodotus focused on the Macrobians' diet, which allegedly consisted exclusively of boiled flesh and milk. While I wouldn't consider those foods Bulletproof, it is fascinating that even back then people had an intuitive awareness that longevity stemmed not just from good genes or good luck, but rather from the environment inside of and around us. And they were willing to make changes to those environments if it meant living longer.

If you've read my other work, you've probably already noticed that the ancient Greeks were biohackers, as were the cavemen before them. When I created the groundswell movement around biohacking, I defined it as changing the environment inside of and around you to gain control of your own biology. (In 2018, Merriam-Webster added *biohacking* to the list of new words in the English language!)

Today there's scientific proof that we can make changes on the sub-cellular level (aka changes that affect the makeup of our cells, including our mitochondria) that will dramatically extend life-span. When I interviewed stem cell biologist Bruce Lipton, he told me that he was able to keep a line of cells alive in his lab for much longer than usual simply by changing the water in their growth medium every day. In other words, he made sure those cells had a clean environment, and as a result they gained longevity. But they did eventually die because one of Bruce's lab assistants fell prey to short-term thinking and forgot to change the water in the growth medium. Maybe he was hungry . . .

If you want to live to a hundred and eighty, or even to an energetic eighty, it's essential to look at your life and ask, "What's going to make me forget to change the (proverbial) water?" The answer is, the messages from your mitochondria telling you to fight, to flee, to feed, and . . . you know. Your mitochondria pay attention to the environment around them, and you can hack that environment so those little guys don't keep you stuck making poor short-term decisions. Unlike Thog or the Macrobians, we now have technology that allows us to change every element of your environment—from your hormones to your nutrition, to the light you're exposed to, to your temperature, to the very vibration of your cells.

Are these "cheats"? No. They are tools we can use to control our biology. And what's the first thing any one of us would do once we gained control of our biology? Not die. The second thing? Age backward. And finally? Heal like a deity so you can keep getting better with age instead of suffering an inevitable decline.

This is exactly what this book will teach you to do. First, we'll look at the biological factors that cause most of the diseases of aging and how you can stop them. Once you've learned how not to die, you'll learn how to age backward with strategies ranging from simple to cutting edge that will add more years to your life and more life to your years. Finally, we'll explore some truly radical anti-aging techniques to help you achieve Super Human status. We tell ourselves that the one thing we can't have more of is time, but that's simply not true. I've seen firsthand how much more life these hacks can give you, both now and in the future.

In case you're wondering, no, we're not going to carefully change one variable at a time while we die waiting to see results. I am an engineer and a biohacker focused on outcome, and I want to feel good now. A research scientist or a medical doctor would approach the problem differently. Scientists get old picking apart every detail to gain a complete understanding of something, a venerable use of time that improves the world. Medical doctors most often focus on treating disease (according to medicine, aging itself isn't a disease) rather than preventing it. However, you are in charge of your own body, and you have the freedom to pick a goal and change multiple factors in your life that might affect the outcome until you get what you're looking for.

Besides, testing out one variable at a time is nearly impossible. If you were to take one supplement for a month to see how it works but you decided to take a different route to work one day, you accidentally changed a variable . . . did that impact the outcome? What about the breakfast you ate or the socks you wore? There are countless variables in our environments that are changing all the time, and I have no interest in keeping track of them all. I want more energy now and for the next hundred and thirty-four years, and I'm willing to change however many variables I need to in order to increase my chances of getting that result.

This is personal for me. Until a decade ago, I never thought I'd make it past eighty, never mind aim for a hundred and eighty. Starting at a young age I was overweight and chronically ill, with arthritis in my knees when I was just fourteen. By the time I was in my twenties I was prediabetic and suffered from brain fog, fatigue, and dozens of other issues we normally associate with aging. My doctors told me I was at a high risk of heart attack or stroke before I was thirty. In short, there was no reason to believe I was going to live a long and/ or healthy life.

Thanks to some wise elders I began working with in the nonprofit anti-aging field, I learned it was possible to prevent additional damage to my cells and even reverse some of the damage that had already occurred. In my late twenties, I decided to invest 20 percent of my net income each year into hacking my biology with nutrition, supplements, lab tests, treatments, technologies, and whatever it took to learn more. There were some years when this was more difficult than others, but there is no higher return on investment than more energy now, likely with additional years of functional life later.

With the help of amazing anti-aging doctors and a community of folks who've been studying longevity since I was in diapers, I was able to take my biology into my own hands. I reversed my diseases and symptoms and began literally aging backward. If I can turn things around after such a poor start, you probably can, too. And the good news is, as these interventions gain popularity and demand for them goes up, the price of them is going down. One of my main goals with this book is to bring these little-known methods out of the shadows of anti-aging circles and into the mainstream so they will become even more accessible.

Not only can you make changes that allow you to live longer than you think possible, but you must. We all have a moral obligation to live well for as long as we can to develop our own wisdom and share it with future generations. By choosing to live longer, you are not taking anything away from anyone. Instead, you are giving yourself an opportunity to share more with the people and the world around you. I see it as our duty to ensure that we are able to share our life experiences, and—just as important—to make them worth sharing.

This, too, is not a new concept. We used to value the wisdom of tribal elders who taught young people how to avoid the mistakes of past generations. If you made it to old age, you were considered a great source of knowledge. But now the people who've lived long enough to develop that wisdom are usually too sick or tired to share it, or else they don't even remember it! This is a crime against humanity. But we can change it.

When you have as much energy at eighty or ninety as you did at twenty-five, you have a tremendous potential to positively impact the world by sharing your wealth of information gleaned from relationships, experiences, successes, and mistakes. If you take that kind of energy and intelligence and put it to work, you can literally improve the world for future generations. Now you're the tribal elder who's leading the hunt because you're full of energy and you've been around a long time, so you know where all the animals are hiding.

Contrary to common fears, our living longer won't lead to overpopulation and environmental ruin. If we use our advanced wisdom and energy to create a world in which everyone had access to a quality education and reproductive health care, we'd actually start to see *negative* population growth.

Americans may struggle to envision a world where we live past a hundred years old, but the governments of countries like China and Russia are investing in anti-aging technologies because they realize that it gives them a tremendous competitive advantage in the world economy. It costs a lot of money to keep reeducating new generations of workers, not to mention caring for a sick and aging population. What if instead of being sick, old people were productive and happy citizens who could contribute to society in their final years?

That is the future I plan on sticking around to see. If you knew it was possible, how would you change your daily decisions and priorities now? In this future, it's not your unborn grandchildren or great-grandchildren who are going to have to deal with the effects of the environmental problems we've created; it's you. Instead of making a mess in your own sandbox, you'd invest in improving that sandbox so you can enjoy it for the unexpectedly large number of years to come.

This is why I am donating a portion of the advance from this book to organizations like the XPRIZE Foundation, which is funding massive initiatives to improve the world's oceans, soil, food supply, and education system, not to mention exploration of space. Thanks to more computer power, more research, and more money going toward fixing the world's biggest problems, change is happening at exponential rates. Whether you know it yet or not, you're part of a race to fix the planet so it supports a population that can live beyond a hundred and eighty. It's up to you to either participate in that race or get out of the way. Go back to your cave if you like, but don't stand in the middle of the road slowing everyone else down.

My goal is to share the techniques with you that have given me the greatest return on my investment of energy, time, and money. It's easy to spend eight hours a day on an anti-aging protocol, but then you're not actually gaining time because you're spending so much of it on these efforts. Instead, I want you to learn how to stop dying, reverse aging, and heal with Super Human speed in the least possible amount of time and with minimal effort.

As you read this book, I hope you'll create your own prioritized list of things to do to live longer and better based on where you are now and where you want to go. Most likely, you won't try everything in this book. And that's fine—it's not a contest. Perfection isn't required. Even I haven't tried out all of these strategies yet (but I'm getting close!).

Yes, some of these technologies are more expensive than others, although many of the most powerful are the least expensive. And while certain interventions are a rich person's game today, that is changing; you can now access a lot of anti-aging technologies for a fraction of what they cost ten years ago, just like the smartphone you have now is far more capable and less expensive than the models that debuted a decade ago. When you start with the most accessible and simplest lifestyle hacks and selectively choose a few affordable technologies to extend your life (or even just your health), you'll buy yourself time so you can afford to wait for the rest to come to you. What could possibly be a better investment than that?

The slope of innovation is steeper than ever, and change is unstoppable. Are you in or are you out? I'm all in. Join me.

PART I

DON'T DIE

Widen your relationship to time, slow it down. Don't see time as an enemy but an ally. It provides you with perspective. Aging doesn't frighten you. Time is your teacher.

—Robert Greene, *The Laws of Human Nature*

THE FOUR KILLERS

THE CURIOUS CASE OF DAVE ASPREY

Until the age of five, I was a normal kid with few health problems. Then my family moved from California to New Mexico, and something in my biology changed. I started acquiring health problems normally reserved for people far older than I was. Today I recognize that my bedroom, which was in the basement of our new house and covered in water-damaged wood paneling (it was the 1970s), was full of toxic black mold. My own home was silently aging me, but nobody, least of all me, was aware of this at the time.

For the next two decades, I suffered from joint pain, muscle pain, asthma, brain fog, extreme emotions, and even weird, frequent nosebleeds. Out of nowhere, my nose would start gushing, and I had unending strep throat that came back every time I finished yet another round of antibiotics. After I got my tonsils out, I started getting chronic sinus infections instead. My body didn't properly maintain blood pressure, so I often got dizzy, and I was easily fatigued.

At the age of fourteen, I was diagnosed with full-blown arthritis in both of my knees. I remember going home after receiving the diagnosis from my doctor, thinking, *How can I have arthritis? That's for old people.* I had always been chubby, but now I was becoming obese. I developed tons of stretch marks, which also disturbed me. Weren't those for pregnant women? I was just a kid!

And can we talk about man boobs? I grew mine when I was sixteen, which would make anyone self-conscious, especially a teenager. The only other guy I knew with a matching set was my grandfather. My hormones were dysfunctional, just like those of my aging relatives. Between the stretch marks and the man boobs, you'd never catch me with a shirt off. The very thought terrified me, and I'd never in a thousand years imagine that thirty years later, there would be a full-page shirtless photo of me in *Men's Health* magazine talking about how I used the techniques in this book to get rid of that flab and replace it with abs.

When I got to college, I kept putting on weight until I had grown a size 46 waist. And my knees got even worse. I played intramural soccer, and my kneecap would become dislocated, so my leg would suddenly fold sideways in a sickening way. I got used to falling over unexpectedly when it happened. Besides the pain, this made dating really awkward. Who wants to date an obese twenty-year-old who might fall down at any moment, with stretch marks, man boobs, arthritis, and the lack of confidence that comes with having such things? Oh, and someone who was so fatigued that he often forgot names, was socially awkward, and could barely focus, even when he really tried? Not too many people, unsurprisingly.

More important than my lackluster social life was the fact that my body was aging before its time. I was well on my way to prematurely developing all four of the diseases most likely to kill you as you age—heart disease, diabetes, Alzheimer's, and cancer—or, as I call them, the Four Killers. These diseases are all deadly, and each of them is on the rise.

Right now, about one in four deaths in the United States is connected to heart disease—that's roughly 610,000 people who die from heart disease each year. Meanwhile, more than 9 percent of the population of the United States has diabetes, and that number rises to 25 percent for people over the age of sixty-five. The Centers for Disease Control and Prevention (CDC) estimates that 5 million Americans are living with Alzheimer's, and this number is going up, too. The death rate due to Alzheimer's disease increased a full 55 percent between 1999 and 2014. And last but not least, 1.73 million people in the United States are diagnosed with cancer each year, and more than 600,000 of them die from it.

Suffice it to say that if you don't die in a car crash or from an opioid addiction, chances are that one of these Four Killers is going to drain your life and your energy (and your retirement fund) before you die in a hospital. It was certainly looking like that would be the case for me—and sooner than most people, given how sick I was.

In the 1990s when I was in my twenties, my doctor used blood tests to determine that I was at a high risk then for developing a heart attack or stroke. My fasting blood sugar was a whopping 117, which put me solidly in the range of prediabetic. I didn't have Alzheimer's, but I was experiencing significant cognitive dysfunction and often left my car keys in the refrigerator. And I may not have been at an obvious risk of cancer, but guess what nearly doubles your risk of certain cancers (including those of the liver and pancreas)? Diabetes[1]—which is also a risk factor for Alzheimer's.[2] Guess what else dramatically raises your cancer risk? Toxic mold exposure, which I had also experienced.

Even obesity itself is the second largest preventable cause of cancer. Your risk goes up the more overweight you are and the longer you stay that way.[3] Bad news—75 percent of American men are obese, and so are 60 percent of women and 30 percent of kids.[4] No wonder the Four Killers are on the rise. Are you going to let them take you out?

I still didn't know what was causing me to age so quickly when I began a quest to discover how to fix my body. In the mid-1990s, we didn't have Google yet, but we had AltaVista, and I worked at night teaching the engineers who were literally building the Internet. This meant I had the good fortune of having access to information that most people didn't. I started doing a ton of research and buying whatever I could find that might help me slow down or even reverse my symptoms. I simply couldn't imagine even more stretch marks or more joint pain as I got older.

An important part of this journey was connecting with one of the first medical doctors who specialized in the study of anti-aging, Dr. Philip Miller. Seeing him required what was a tremendous financial investment for me at the time, but I was desperate. My first visit with Dr. Miller was like nothing I'd ever experienced. He ran new kinds of lab tests that regular doctors at the time didn't know existed, including the first real hormone workup I'd ever had. Then he sat me down and gave me the bad news: I had Hashimoto's thyroiditis (an autoimmune condition that causes the body to attack the thyroid) and almost no thyroid hormones, and my testosterone levels were lower than my mom's. (He had done a workup for my mom not long before, so he wasn't exaggerating when he told me this.)

The news could have been devastating, but I was actually excited to have the hard data. I felt in control for the first time because I finally had real information and knew exactly what I needed to change. This was proof that it wasn't just a deficiency in my effort or some sort of moral failing. It's common to see your hormone levels drop off around middle age, but not in your twenties. Now I had proof that I was aging prematurely and not just lazy, and I was determined to turn things around.

Dr. Miller and I came up with a plan for me to restore my hormone levels to that of a young man using bioidentical hormones and continue to track my data. The hormones made an enormous difference right away. I got my energy back along with my zest for life. It gave me so much hope to know that I could actually reverse some of my health issues, which I now knew were common symptoms of aging. So when I heard about an anti-aging nonprofit group in Silicon Valley, now called the Silicon Valley Health Institute (SVHI), I decided to check it out.

As I sat there at the first SVHI meeting listening to people who were at least triple my age, I felt completely at home. These were my people, I realized. I had more in common with them than I did with most of my peers, except these people had decades of wisdom I didn't. After the meeting, I stayed for a long time talking with a board member who at eighty-five years old was kicking ass and full of energy in a way that was amazing and seemed totally impossible to me—but that I was inspired to replicate.

For the next four years I focused completely on learning as much as I could about the human body. I studied medical literature, read thousands of studies, talked to researchers, and spent all my free time at SVHI learning from seniors who were actively reversing their own symptoms of aging. This completely changed the way I thought about health, as well as aging. I learned that there is no one thing that causes disease or that leads us to age. Instead, aging is death by a thousand cuts, the cumulative damage caused by little insults stemming mostly from our environment.

Then in the year 2000 I found a former Johns Hopkins surgeon who ordered a litany of tests, including some allergy tests that showed I was highly allergic to the eight most common types of toxic mold. That was the smoking gun. In order for my immune system to be sensitized to those toxic molds, I must have been exposed to high levels of them, which wreaked havoc on my cells. This was one of the unexplained environmental factors that had made me age so rapidly.

My premature aging makes complete sense to me now. Mitochondria, which are bacteria embedded in most of our cells, power our energy production. Back when we were single-celled creatures, we became host cells for ingested bacteria. Over millions of years of evolution, the host cell became humans, the ingested bacteria became mitochondria, and today neither of us can survive without the other. Mitochondria are not of human origin; they even have their own DNA. And what has posed a lethal threat to bacteria since the beginning of time? Mold.

This means the very powerhouses of my cells were constantly engaged in a battle with their mortal enemy, and this fight left behind many casualties. When cells are under chronic stress, their mitochondria cannot make energy efficiently. This leads to an increase in the production of molecules called reactive oxygen species (ROSs), also known as free radicals. ROSs are unstable molecules that contain atoms with unpaired electrons, making them highly reactive. When an excess of free radicals are present in cells, they cause a chemical reaction that damages your cellular structures in a process called oxidation.

This is *exactly* what happens as you age, whether or not toxic mold is present in your life: Mitochondria function steadily declines, leading to an increase in free radicals, which damage your cells. In response, your body sends vitamin C from food to the liver so it can produce antioxidants, which fight off free radicals. The problem with this process is that it leaves you without enough vitamin C to produce collagen, the protein in the connective tissue of your skin, teeth, bones, organs, and cartilage. Vitamin C interacts with amino acids to build collagen, but only if you have enough of it. Your body will gladly sacrifice healthy blood vessels and skin in favor of fighting off free radicals that are draining its energy source.

This is precisely why I had stretch marks and vascular issues (manifested as nosebleeds) and why most people don't develop these symptoms until they're much older. The fight in my body between my onboard bacteria and mold left me constantly depleted of antioxidants. And my mold-damaged mitochondria also laid the groundwork for prediabetes, poor blood flow to the brain, arthritis, cognitive dysfunction, and, according to one doctor, a high risk of stroke and heart attack. I was still in my twenties, but I was biologically *old* because my mitochondria were slowing down. And it really pissed me off.

MITOCHONDRIA AND THE FOUR KILLERS

As I fought my way back from experiencing the many symptoms of aging, my likelihood of dying from the Four Killers dropped dramatically. That's because—surprise, surprise—they all have one underlying issue in common: the cumulative damage to your cells, and in particular, to your mitochondria, that takes place over the course of a lifetime. This damage occurs in all of us, though at varying rates. Some damage stems from the bad choices we make, but much of it is simply the price we pay for the basic functions that support life—like metabolizing food and breathing.

You die a little bit every day from these cuts that make you weaker in the short term and hasten your decline in the long term. Staying alive requires avoiding as many of those cuts as possible, but they are all around you—in your food, your air, your light sources, and throughout your environment. You may not associate these cuts with your likelihood of aging prematurely or of developing a degenerative disease, but like every other aspect of your biology, they are all connected. The cuts lead to aging, aging leads to disease, and disease leads to death.

If you're in your twenties or thirties, you may think you're in the clear—that these cumulative cuts aren't affecting you yet. But the cuts from bad choices or a toxic environment begin to add up from an early age—and they're hurting you even if you're not currently feeling their effects (such as weight gain, brain fog, muffin top, and fatigue).

And it's a lot easier to avoid damage to your mitochondria than it is to reverse it later.

Your mitochondria are responsible for extracting energy from the food you eat, and then combining it with oxygen to produce a chemical called adenosine triphosphate (ATP), which stores the energy your cells need to function. When your mitochondria conduct this process efficiently, they produce lots of energy so you can perform at your greatest potential—like a young person. But if your mitochondria become damaged or dysfunctional as you age, they begin producing an excess of free radicals in the process, which leak into the surrounding cells and lay the groundwork for the Four Killers. Congratulations, you are now old.

Even young, efficient mitochondria produce some free radicals as by-products of creating ATP, but they also make antioxidants, compounds that inhibit the damaging effects of free radicals. This is why products containing antioxidants have "anti-aging" properties. While popping antioxidant supplements and using skin-care products containing antioxidant-rich ingredients are worthwhile interventions, they are, frankly, the low-hanging fruit of our Super Human tree. For you to truly remain young, those antioxidants have to be produced by your body—your mitochondria must create at least as many of them as it does free radicals. When your mitochondria become inefficient, they make an excess of free radicals and fewer antioxidants. And you can't slather enough serums onto your skin to fully counteract the damage created by this imbalance.

Your mitochondria are also in charge of triggering cellular apoptosis, programmed cell death that occurs when a cell is old and/or dysfunctional. If your mitochondria are sluggish, they may not trigger apoptosis at the right times, which can result in healthy cells dying off before they should or dysfunctional cells sticking around past their prime and aging *you* before your time.

When you're still young and exploding with mitochondrial energy, you can take some of these hits. You can eat garbage, drink too much cheap beer, forgo sleep, and still function pretty well because you're producing lots of antioxidants and energy. As you get older, you start to see that you can't stay out all night drinking and still really bring it at work the next day. By the time you wake up to this new reality, you've already taken a lot of hits that will age you in the long run. But you're likely to keep running at the edge of what you can perceive, so the damage stacks up without you even knowing it.

Well, what if you made better choices throughout your life so you took fewer hits over the course of decades? Then when you got to the age of seventy you might look and feel more like fifty because you simply suffered less damage. You're never going to be able to avoid all the cuts—again, simply breathing creates some amount of wear and tear over time. It's a matter of preventing as much damage as possible, which happens to dovetail nicely with the first rule of biohacking: Remove the things that make you weak. This is in and of itself a powerful anti-aging strategy.

When your mitochondria start to slow down and create an excess of free radicals, the result is widespread chronic inflammation throughout your body. Inflammation is such a hot topic in the field of longevity that you probably already know how closely it's linked to aging. When I was sick and old as a young man, I knew I was inflamed, but I had no clue this stemmed from mitochondrial dysfunction, nor did I know that inflammation was more than a painful annoyance. I had no idea that inflammation creates the ideal circumstances for each of the Four Killers to thrive.

HEART DISEASE

A condition known as atherosclerosis, hardening of the arteries, is the first obvious clinical sign that heart disease has started. But what causes this? A thin layer of cells called the endothelium lines your arteries. When the endothelium is damaged, fats can cross into the arterial wall and form plaques. This is bad enough, but when your immune system picks up on the fact that this is happening, it creates chemical messengers called inflammatory cytokines to attract white blood cells to those plaques. This is an inflammatory immune response. When those plaques rupture because they are so inflamed, blood clots form, and these clots are the real cause of most heart attacks and strokes.

While some doctors are hesitant to definitively state that inflammation causes heart disease, it's hard to refute the evidence that inflammation is a big step in the disease's process, and most functional medicine practitioners now identify inflammation as a bigger health risk than cholesterol levels. In a landmark study conducted by researchers at Brigham and Women's Hospital that followed ten thousand participants for twenty-five years, the data revealed that participants who reduced their inflammation levels also significantly lowered their risk of cardiovascular disease and the need for heart surgery without any other medical interventions.[5]

A new study out of the University of Colorado at Boulder shows that your gut bacteria actually play a role in the inflammation behind atherosclerosis.[6] As animals (and likely humans) age, changes to gut bacteria harm the vascular system and make arteries stiffer. That stiffening came from inflammation. The gut bacteria of older mice actually produced three times the normal amount of an inflammatory compound called trimethylamine N-oxide (TMAO). When researchers used antibiotics to knock out the old mice's gut bacteria, their vascular systems magically returned to those of young mice. The researchers concluded, "The fountain of youth may actually lie in the gut." After following the lifestyle recommendations in this book, I am happy to report that my last test showed that I had zero species of gut bacteria that produce this harmful compound!

Even more mind-blowing, a 2017 study out of the University of Connecticut in Storrs revealed that the fat molecules that form plaques in your arteries come not from the fat in the food you eat, but directly from bad gut bacteria.[7] This turns everything that conventional doctors tell us about dietary cholesterol on its head and means you have permission to laugh when people repeat the myth that a "plant-based" diet is better because it doesn't contain saturated fats like butter that will somehow "stick to" your arteries. It also shows the importance of healthy gut bacteria and mitochondria for a long and energetic life. (More on this in chapter 11.)

We know that the mitochondria in our cells, which themselves evolved from bacteria, communicate with the bacteria in our gut. Bacteria communicate with one another via chemicals (like hormones), light, or physical movement. They even gather around and trade bits of their genetic code in a microscopic swap meet for bacteria superpowers. This is called a plasmid level exchange. Imagine a group of Marvel superheroes hanging out at headquarters. Wolverine says to Spider-Man, "Do you want my ability to grow claws? I'll trade you for your super speed." This happens constantly in our guts and in the world around us, which is why drug-resistant bacteria spread so rapidly. It's also why we must end industrial livestock practices that require antibiotics. The bad bacteria that evolve in that environment find their way into your gut and make it hard for you to live well for a long time.

So there is clearly an inflammatory and gut bacterial connection to heart disease. Plus, we know that when you have the right kind of bacteria in your gut they can actually transform the foods you eat into short-chain fatty acids, which are highly *anti*-inflammatory. Nurturing healthy gut bacteria is one of the most important things you can do to become Super Human, and you'll learn how later.

Look, I remember what it felt like when my doctor, complete with white lab coat, looked right at me and said in a matter-of-fact voice, "You are at a high risk for heart attack and stroke." I recall the bewilderment and fear in my gut as I stared my own mortality in the face. That happened when I was still in my twenties, and thanks to the information in this book, it is not an issue for me anymore. But even when I was just a kid, I had symptoms of cardiovascular issues, specifically blood pressure instability, a condition normally reserved for much older people. When I stood up quickly, my blood pressure was too low to keep oxygen in my brain. This caused me to start seeing stars and feel extremely fatigued. As a youngster, I would lean my head forward after getting out of a car in order to avoid seeing stars. I was so used to this that I thought it was how everybody lived.

Now I know these were symptoms of postural orthostatic tachycardia syndrome, or POTS, which is often triggered by toxic mold exposure but can also happen with age. In either case, inflammation disrupts the line of communication between the nervous system and the endocrine (hormonal) system. The disruption of these signals leads to fatigue and blood pressure instability, and can lead to symptoms of attention deficit disorder (ADD)[8] and Asperger's syndrome,[9] which I certainly exhibited as well.

This manifested in my not knowing the names of most of the kids in my class, even at the end of the school year. I had zero facial recognition and no understanding of basic social skills. My body was filtering out those signals to conserve energy because my biology was so trashed. Our bodies will always prioritize survival over socialization, and I didn't have enough energy to go around.

It may be hard to comprehend how cognitive symptoms could be connected to vascular issues, but as you will learn in this book, everything in the body is connected. And that includes the diseases that age us and too often lead to premature death.

DIABETES

While the idea of inflammation "causing" heart disease remains controversial, we have definitive proof that type 2 diabetes is an inflammatory disease,[10] and having diabetes dramatically increases your risk of cardiovascular issues. More than ten years ago, researchers discovered that when macrophages—immature white blood cells that play a key role in the immune response—find their way into otherwise healthy tissues, they release inflammatory substances called cytokines that cause nearby cells to become insulin resistant.[11]

In insulin resistance, the body has an impaired response to insulin, which is normally responsible for moving sugar out of the blood and into your cells. The result is that your blood sugar levels are not well regulated and become chronically high. Because chronic high blood sugar will eventually lead to diabetes—a disease in which the pancreas is unable to produce enough insulin to keep up with the body's demands—a diagnosis of insulin resistance is most often accompanied by the label *prediabetic*. Prediabetes is so common now that it almost seems like no big deal. The CDC says that more than one out of every three Americans is prediabetic. But it is actually a huge deal because having diabetes dramatically increases your risk of developing the other killers.

Excess blood sugar causes damage to the entire vasculature, so if you have diabetes, you're more likely to have heart disease or a stroke. High blood sugar also causes dangerous nerve damage by injuring the walls of the capillaries that bring blood and nutrients to your nerves.

This is called peripheral artery disease, and it is especially common in the legs and feet, which is why you may have heard of people suffering from diabetes needing foot or leg amputations. When this happens in the eyes, it causes blindness. If that's not bad enough, diabetes can damage your kidneys' filtering system, resulting in kidney disease. And finally, the higher your blood sugar, the greater your risk for Alzheimer's disease, to the point that some researchers call Alzheimer's "type 3 diabetes." So you've got to keep your blood sugar levels stable, no matter what.

You may think you're off the hook if you are not overweight, but you can be thin and still be prediabetic (or even fully diabetic). Those problematic macrophages are most likely to trigger inflammation in adipose tissue, aka fat. So the more excess fat you're carrying, the higher your chance of becoming insulin resistant and developing type 2 diabetes. But the same thing can happen if you are not overweight but have excess visceral fat, which is the type of fat that's packed around your internal organs instead of underneath the skin. This "skinny fat" is even more dangerous than fat you can see.

There is new evidence that maintaining normal amounts of muscle strength as you age can help ward off this killer. In a study following five thousand people for over twenty-five years, participants were given regular strength tests. The risk of diabetes was slashed by 32 percent in those with even moderate muscle strength as opposed to those with low muscle strength.[12] The reduced risk did not change if the participants were even stronger, so you don't have to get ripped to live longer, but you should avoid carrying excess fat.

I had no idea as an obese teenager that inflammation was making it difficult for me to control my blood sugar. Instead, I bought into the myth that I just wasn't trying hard enough to lose weight. I exercised a ton and constantly watched what I ate. For breakfast, I had Grape-Nuts, which were supposed to give me energy, and skim milk, which was meant to do my body good. But they did neither of those things. I distinctly remember one morning in ninth grade eating a bowl of Grape-Nuts with skim milk to prepare for a big soccer match. I was convinced this was a healthy breakfast, but I didn't perform very well in the game. I thought to myself afterward, *Well, that didn't work the way it was supposed to.*

This was the first time I questioned conventional wisdom about what was actually good for me. It would be many more years before I started to get real answers, but in my desperation I started experimenting with things that no teenager should need to explore. I was sick of feeling like an old man. So I started reading everything I could get my hands on that offered some advice for how to feel and perform better. While my peers were (I assume) out drinking and having fun, I was at home biohacking.

For my knee pain, I tried the glucosamine pills from the health food store, and they brought some serious relief. I didn't know it then, but glucosamine inhibits glycolysis, your body's breakdown of glucose (sugar). As a result, your body has to get energy from fat instead of sugar, which helps prevent insulin resistance. Recent research on mice has found that glucosamine promotes mitochondrial biogenesis (the birth of new mitochondria) and mimics the effects of calorie restriction.[13] And there are plenty of studies to show that calorie restriction (a diet consisting of fewer than 1,200 calories a day) in conjunction with good nutrition extends life-span. In mice, calorie restriction can extend life-span by as much as 40 percent. Most researchers estimate that the impact on humans is more like 10 percent, which is still pretty amazing[14]—if you're willing to be hungry, anyway.

If you're like most people, you don't enjoy feeling hungry, and you don't want to restrict your calories to fewer than 1,200 a day. The good news is that researchers have been testing compounds that mimic the benefits of calorie restriction without the starvation. Glucosamine is one of those compounds. In one study, glucosamine extended the life-span of mice by 10 percent.[15] And it most likely helped with my knee pain because of the way it impacted my body's sugar metabolism.

Despite this small win, I was heavier than ever and fed up. In college I spent eighteen months working out six days a week for an hour and a half at a time while on a low-calorie, low-fat semi-vegetarian diet with lots of rice and beans and everything that was supposed to be good for me. I got really strong, but I was still covered in blubber, and later blood tests revealed that I was prediabetic thanks to all that fat and the inflammation it was fueling.

I knew something had to change, but I had no idea what that thing was. Then one day while I was at a coffee shop getting my daily fix,

I spotted a weightlifting magazine on a rack. No one I knew in my small farming town read weightlifting magazines, but something on the cover caught my eye. It said, "How to grow abs!" Looking down at what I had grown—which you could more accurately call *flabs*—I thought, *I have to read this.* The goal seemed impossible in the world I lived in.

As I sipped a triple latte, I read an article by a body builder with impressive abs who said that sugar and carbohydrates make you fat. That advice was radical at the time and is still mildly controversial today, but it has become much more widely accepted since we know for a fact that sugar causes inflammation.[16] Even small spikes of blood sugar have a particularly bad effect on your vascular system (and raise your cancer risk, too).[17] I grabbed the magazine, went home, and made a smoothie out of cottage cheese and orange juice. I had no idea what I was doing! But that gross smoothie still had fewer carbs than I was used to eating in my efforts to get healthy.

I started eating more protein and avoiding grains and most obvious sources of sugar, for the first time focusing more on what I *didn't* eat (carbs) than *how much* I ate. In three months I lost fifty pounds, but more surprising were the changes to my personality. Everyone in my life noticed that I was a lot nicer, and I actually started to develop friendships. I had changed my biology enough that I wasn't exhausted all the time, and my brain was able to learn how to connect with people, even though it still didn't come naturally to me. My focus in class also improved, and my GPA went up dramatically, from a 2.8 the previous semester to a double course load with a 3.9.

That's right—avoiding grains and sugar helped me reduce inflammation, stabilize my blood sugar, get smarter, and change my personality for the better. Once again, everything is connected. Realizing I'd been fed a bunch of lies (literally) about what to eat for most of my life, I dug into the research and tried different strategies, evolving from the cottage cheese smoothie to the Zone diet to Atkins. (Though I never got anywhere close to the abs promised in that magazine.) Eventually I realized that there had to be a science to this. There were clearly foods out there that acted as kryptonite, caused inflammation, and completely threw me off of my game. And when I ate them, I not only felt awful, but I was also one step closer to developing type 2 diabetes. It took me years, but I finally discovered what those inflammatory foods were and how to avoid them. You'll read more about this in chapter 3.

ALZHEIMER'S

Just as immune cells in your body fat create inflammation that contributes to diabetes, there are specialized immune cells in the brain called microglia that perform similar functions. They control the brain's immune and inflammatory response and are also in charge of killing off dysfunctional neurons in a process similar to apoptosis. The microglial cells constantly monitor the brain, and when they sense a threat, they trigger the release of inflammatory cytokines to attack and remove potential pathogens. This process causes inflammation, and if it becomes chronic this can damage or kill neurons, causing memory loss and other cognitive problems.[18] Many researchers now believe that this is the root of Alzheimer's.

In my twenties I was already experiencing significant cognitive dysfunction, and I wondered in the back of my mind if I was on track for developing Alzheimer's. When I was in business school in the 1990s, my performance on tests was horrible. On exams with math questions, my grades showed a linear decline in per-question scores—100 percent on the first question, 70 percent on the next, 30 percent on the next, and directly downhill from there. My brain got fatigued so easily, even when I studied and knew the answers.

This experience led me to imagine what would happen if I couldn't rely on my brain to earn a living. I'd had a successful career so far, but suddenly I wondered if I wasn't as smart as I thought I was. I decided to undergo a then controversial brain imaging technique called a SPECT scan to see what was really going on in my brain. It showed that my prefrontal cortex—the part of the brain involved in complex thinking and decision-making—had essentially no activity when I tried to concentrate. Dr. Daniel Amen, who was one of the first people in this country to use SPECT scans, was shocked that I had been even remotely successful in my career with such clear cognitive dysfunction.

Once again, receiving bad news actually came as a relief. It was incredibly validating to hear that there was indeed a reason why everything felt like such a struggle. The issue wasn't lack of effort or intelligence. It was an actual biological problem, a hardware problem. And there were lots of little-known things I could do to reduce inflammation and improve my brain function. When I found these interventions, the impact was immediate and allowed me to get smarter and faster with each passing year. The good news is that once you know them, the interventions are simple and practical.

If you're in your twenties or thirties, it is much easier to reduce inflammation now to boost your brainpower and avoid cognitive decline with age, but even if you are older or experiencing symptoms of dementia, it is still possible to improve your brain function. The sooner you start, the better, but it's never too late to begin growing a younger, more powerful, and more energetic brain. You'll learn how to do this later in this book.

CANCER

More than 40 percent of Americans are diagnosed with cancer in their lifetime.[19] When mitochondria become dysfunctional and don't produce energy efficiently—which, again, is typical of most people as they age—your risk of cancer increases. This is because an inflamed environment offers the perfect conditions for cancer cells to proliferate.

Think about a time you got a cut and the wound became swollen—an obvious sign of inflammation (an immune response) at work. When the body is injured, your cells multiply quickly so the wound can heal. That process alone does not cause cancer. But when cells multiply rapidly in an environment that contains excess free radicals—which damage the DNA of cells—the risk is that damaged or mutated cells will proliferate. If these damaged cells continue to reproduce, the result can be cancer.[20]

We often think that our risk of developing cancer is based mostly on our genetics, but the data shows that only about 2 to 5 percent of cancers are truly genetically based, and mitochondrial dysfunction causes most others. In 1931, a German biochemist named Otto Warburg won the Nobel Prize for discovering that highly dysfunctional mitochondria actually stop burning oxygen to make energy and turn instead to a much less efficient process called anaerobic metabolism, which is the combustion of carbohydrates in the absence of oxygen. Anaerobic metabolism is associated with the vast majority of cancers. But if your mitochondria are strong, they will not have to resort to anaerobic metabolism. This greatly reduces your cancer risk.

Cancer is something of a double-edged sword when it comes to anti-aging. Any time you do something that makes your cells grow faster or get younger, you are inherently increasing your cancer risk because cancer cells can potentially grow and rejuvenate along with the healthy ones. Then you end up with this weird dichotomy: You can grow old "normally" with a roughly 40 percent chance of getting cancer, or you can get younger and maybe as a result slightly increase your risk. My solution to this dilemma is to do everything I can to make sure my mitochondria run like superstars because that in and of itself will reduce my risk of cancer. I also take action to promote my body's natural detoxification efforts.

In addition to apoptosis, which you read earlier is healthy, controlled cellular death that targets old or unstable cells, your body also has a built-in detox process to recycle damaged cellular components. This is called autophagy, a Greek word that translates as "self-eating." During autophagy, your cells scan the body for pieces of dead, diseased, or worn-out cells, remove any useful components from these old cells, and then use the remaining molecules to either make energy or create parts for new cells. This recycling process removes unwanted toxins, reduces inflammation, and helps to slow the aging process.

When you activate autophagy, you slow down the aging process, reduce inflammation, reduce your cancer risk, and increase your body's ability to function at its best. There are specific supplements and lifestyle modifications such as brief bouts of fasting that boost autophagy. You'll learn how to do this as we get deeper into the techniques that make you Super Human.

SLASH YOUR RISK

Despite overwhelming evidence that mitochondrial dysfunction and the resulting inflammation leads to the Four Killers, we live in a society in which an inevitable decline in mitochondrial function is considered a normal part of aging. Of course we expect to die from one of these diseases! Between the ages of thirty and seventy, you experience a decrease in efficiency of the average mitochondrion by about 50 percent, setting the stage for you to develop these killers.

Since you're reading this book, you obviously have no intention of aging like an average person, and you shouldn't. By the time I discovered the importance of mitochondria, mine were already trashed from years of toxic mold exposure. The mold had weakened my system and aged me prematurely, so in many ways I was the canary in the coal mine. I felt the "cuts" that affect all of us much sooner than most people because I started off in a weaker spot. In order to get to a basic level of functionality, I had to find out what was causing these cuts and work on eliminating them.

Feeling the cuts so early and so deeply allowed me to experience real-time feedback and determine which environmental factors impacted my health and performance the most. This turned out to be an enormous gift because I was able to learn—and can now teach you—how to stop damaging your own body with thousands of invisible cuts by focusing on the basics: good nutrition, quality sleep, and a healthy environment free of toxins that cause more cuts.

Before we move on to learn how to do that, let's take a closer look at exactly what these cuts do to our bodies. Obviously, you won't go from eating an inflammatory meal to developing degenerative disease in one fell swoop. Instead, the cuts from your environment cause invisible damage on the subcellular level. This damage doesn't age you at once, but it does so cumulatively, day after day, year after year. By the time you become aware of this damage, you're old. But you can take action now to stop this damage before it stacks up. So after you take the steps to avoid the Four Killers, it's time to focus on cheating death the way Super Humans do—by avoiding the Seven Pillars of Aging. These are the processes in your body that break as you age, and there's a lot you can do to control them.

THE FOUR KILLERS

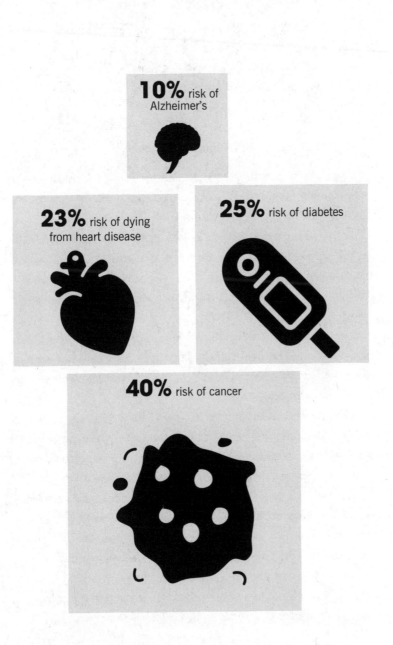

10% risk of Alzheimer's

23% risk of dying from heart disease

25% risk of diabetes

40% risk of cancer

Bottom Line

If you are average . . .

- You have a 23 percent risk of dying from heart disease.
- You have a 25 percent risk of diabetes.
- You have a 10 percent risk of developing Alzheimer's.
- You have a 40 percent risk of cancer and a 20 percent risk of dying from it.

So start hacking. Do these things right now:

- If you have joint pain or blood sugar issues, consider taking glucosamine, which helps control blood sugar and extends the lifespan of mice (and probably humans).
- Consume more antioxidants to fight off free radicals. Berries, herbs, spices, coffee, tea, and dark chocolate are good sources. There are also medical spas in most cities that offer antioxidant therapy via IV. It may be worth looking into if you travel frequently or need an energy boost.
- Short periods of fasting stimulate autophagy. You'll read more about the longevity benefits of fasting and how to do it without hunger later, but it's worth starting now to benefit right away from increased autophagy.
- To help with cardiovascular issues, try the Zona Plus, a digitally controlled handheld device that uses the science behind isometric exercise to increase both vascular flexibility (thus decreasing blood pressure) and the production and flow of nitric oxide throughout the body, which is linked to treating various cardiovascular conditions, erectile dysfunction, and muscle fatigue. It's a cool biohack for anyone who wants to improve their cardiovascular health.
- While it is most useful to look at how the environment will control your energy levels and your aging, it's not like your DNA is meaningless. The area of functional genomics is just getting going. Like functional medicine, it is the study of what you can actually do to

influence risk besides worry about it. For instance, a functional re-view of my genome from the DNA Company revealed that I should take extra steps to take care of the tight membranes in my arteries, including taking the supplements in this book. Check out their tests to discover your weaknesses and learn how to combat them.

THE SEVEN PILLARS OF AGING

Okay, now you've decided not to let the Four Killers take you out. That means it's time to shore up the Seven Pillars of Aging. When I was working to reverse my early aging as a young man, I learned that there are specific forms of cellular aging that drive all forms of aging and disease, even my premature symptoms of aging. Later, I gleaned more detail from longevity experts such as Aubrey de Grey, who is the chief science officer at the SENS (Strategies for Engineered Negligible Senescence) Research Foundation, which has the ambitious mission of curing aging by funding anti-aging research around the world. A lot of my elite anti-aging longevity friends (yes, I have weird and awesome friends) are focused on what SENS calls the "classes of cellular and molecular damage that constitute aging." I call them the Seven Pillars of Aging.

THE SEVEN PILLARS OF AGING

It's important to understand how the Seven Pillars of Aging affect your body on the cellular level. While some degeneration over time is a given, there's a lot you can do to protect yourself from the worst of it. From simple and inexpensive lifestyle changes and nutritional modifications to high-end technologies that are rapidly becoming more affordable, I'll outline multiple strategies to hack the aging process—most of which I've tried out myself.

Is this anti-aging science still nascent? Yes. Do we have ironclad evidence that these strategies work perfectly? No. But we do have some pretty compelling research that suggests they will help you spend more quality years on this planet. Plus, they're not going to kill you—and aging definitely will. So why not give them a shot?

First, let's take a closer look at each of the aging pathways and how they affect us.

PILLAR 1—SHRINKING TISSUES

When you are young, your body has a multitude of stem cells—undifferentiated cells that are capable of giving rise to many more cells of the same type. When cells die via apoptosis, your stem cells spring into action to replace them. As you age, however, a few things happen. Your stem cell reserves dwindle, your stem cells themselves age and thus become less efficient at replacing dead cells, and your mitochondria may not trigger apoptosis at the right times. Some cells die before they're supposed to. Others aren't quickly replaced. As a result, tissues throughout your body lose more and more cells and begin to atrophy, or break down.

Quick, picture a stereotypical "old person." In your mind's eye you probably see a frail person with loose skin, no muscle tone, shaky hands, and a foggy memory—right? The truth is that these things happen as we age and cells die and are not replaced. In fact, loss of muscle tissue is so common that it has its own name, *sarcopenia*, a condition that can lead to falls and broken bones and even impairs the body from fully recovering after those tumbles (or a surgery).[1] In most people, sarcopenia sets in as early as age thirty and gets worse with each passing decade.[2]

When neurons in the brain die and your body isn't able to replace them, your brain literally shrinks. And yes, this typically happens as we age. This contributes to cognitive decline and dementia, as well as a decrease in fine motor skills. In particular, when that neuron loss takes place in the hippocampus—the part of the brain that controls emotion, memory, and the nervous system—you begin to look and sound a lot like that old person you just imagined. Since hippocampal atrophy is so common, the size of your hippocampus is considered a key marker of aging.[3] But there is nothing normal about it—at least, there shouldn't be.

So the big question becomes, What can you do to make sure those dead cells get replaced (or don't die in the first place)?

It turns out that if you keep your mitochondria healthy, you can avoid a lot of unnecessary cell loss. The biggest game changer here is eating foods that boost the efficiency of your mitochondria so they can make more energy and your body has the raw goods it needs to manufacture all the proteins, hormones, and fatty acids they require to function. We'll cover those foods in the next chapter.

It is possible to reverse atrophy by taking advantage of stem cell therapy—a medical treatment in which stem cells are injected into your body. I've had more stem cell treatments than probably anyone on Earth (more on this later), and it's been a game changer for me. I went from having chemical-toxin-induced brain damage as a young man to having a hippocampal volume that's in the 87th percentile for someone my age. But stem cell therapy does not come cheaply or easily, so it's better to prevent atrophy in the first place!

The key to success with any of these interventions is to start them now, even if you don't think you need them. After all, humans are good at avoiding things that hurt. You don't step on nails or burn yourself because you feel the impact of those cuts immediately. But when it comes to aging, you're the proverbial frog in a pot of slowly heating water. You keep taking the hits because you don't feel the impact right away. But cutting down on just a few of those hits by making small changes in your environment can really ramp up what's possible for you. Perfection not required.

PILLAR 2—MITOCHONDRIAL MUTATIONS

Mitochondrial mutations—aka damaged mitochondria—are the second pillar of aging. The importance of this aging pathway cannot be overstated. When the power plants of your cells—the very things that create the energy that keeps you alive—start mutating, is it any wonder everything goes haywire?

Unfortunately, this is a cause of aging that is often overlooked. Even those on the forefront of biotechnology have been so focused over the past several decades on mapping the human genome that they have paid little attention to mutations in mitochondrial DNA. Don't get me wrong—the sequencing of the human genome changed the world, and I'm grateful for the scientists who accomplished this monumental task. But unless you have a significant genetic disorder, it turns out that the human genome is not usually of great value in predicting how you'll age compared to your mitochondrial DNA status.

You can think of your genetic code as the building plans for your body—but who wants a body that doesn't have any energy? (Hint: That's called death.) Remember, your mitochondrial genome is separate from your human genome—mitochondria evolved from bacteria and have their own genetic code. But your mitochondrial DNA serves a very important function—it controls how your body makes energy. Unfortunately, your mitochondrial DNA is a lot more susceptible to mutations than your human DNA because mitochondrial DNA has a limited ability to repair itself when it is damaged. So you're going to want to take fewer hits to your mitochondria.

Think of it like this: Your DNA provides a picture of how a building (your body) will look—how many rooms, how many windows, what kind of roof, how tall, etc. Your mitochondrial DNA describes what kind of wiring, heating, lighting, and air conditioning will be in the building. The building itself is going to be there for a while, but if the wiring goes bad, the air conditioner breaks, and the bulbs burn out, it's not going to be a building you want to live in. Mitochondrial DNA breaks and mutates easily, which is why it's so important.

When it comes to aging, it's helpful to look at epigenetics, the science of how the environment in and around our bodies influences genetic expression, and how these changes are passed down from generation to generation. A 2018 review from leading stem cell researchers found that mitochondrial epigenetic mechanisms influence cell fate, cell division, cell cycle, physiological homeostasis, and even pathologies.[4] In other words, your ancestors' environment controls the mitochondria in *your* cells. And of course problems with those important cell functions can lead to every one of the Four Killers.

As you read in chapter 1, mitochondrial DNA can become damaged when excess free radicals are present. Damage to mitochondrial DNA from free radicals causes deletions in the mitochondria's genetic code. The damaged mitochondria produce energy inefficiently, generating huge amounts of additional free radicals and less of the energy that you would rather put toward your Super Human efforts. And as we know, damaged mitochondria generate inflammation and accelerate aging throughout the body.

Remember, this cycle all started with an excess of free radicals, which are created by . . . dysfunctional mitochondria! So the more efficient your mitochondria, the less likely you are to suffer damage to your mitochondrial DNA, regardless of the genetic code you inherited from your parents. This is one of many reasons my anti-aging protocol focuses so heavily on making sure my mitochondria are running like rock stars for the long term.

PILLAR 3—ZOMBIE CELLS

Death-resistant cells, aka senescent cells, are cells that won't die when they're worn out, and they are a major focus of anti-aging research today. These cells no longer divide or function properly. These cells literally become dead weight. They do not function, yet they persist and secrete inflammatory proteins, causing all the problems that stem from chronic inflammation,[5] including an increased risk of the Four Killers. Even worse, the mitochondria in senescent cells become dysfunctional and release huge amounts of reactive oxygen species. This is called senescence-associated mitochondrial dysfunction (SAMD), and it will age your body very quickly.[6]

Over time, you gain more and more senescent cells, and the accumulation of the damage they create is a major cause of aging and disease. For one thing, when you have too many zombie cells in your tissues, your body becomes less efficient at responding to the hormone insulin. This is the definition of insulin resistance, which as we discussed earlier is a precursor to type 2 diabetes. Zombie cells also lead to an increase in visceral body fat, the type of fat that is stored around important organs in the abdominal cavity and is associated with an increased risk of many diseases, particularly type 2 diabetes.

Zombie cells also contribute to many symptoms of aging that won't kill you but will make your later years a lot less comfortable. For instance, doctors have known for years that patients who need knee transplants have excess senescent cells in their knee cartilage. In fact, injecting just a few senescent cells into your knees can actually cause arthritis.[7] Did senescent cells in my joints cause my arthritis when I was fourteen? Possibly.

Some senescent cells are easy enough to kill off. Others persist like a Netflix binge of *The Walking Dead*. Perhaps the most damaging types of zombie cells are immune cells. Remember, when you get a cut or an infection, your immune cells proliferate to promote rapid healing. Once you have healed, those extra immune cells are supposed to die off. When they don't die, they inhibit your immune system's ability to respond to new infections or injuries. This is one reason the immune system usually becomes weaker as we age. Talk about not dying— pneumonia and the flu together are the eighth leading cause of death in the United States and are both much more common and deadly in people over sixty-five. This is in part because of senescent cells weakening the immune system.

The good news is that there are many things you can do to prevent damage from zombie cells. One of the most important is to keep your cell membranes strong so the cells can function well for as long as possible. I take a supplement comprising calcium, magnesium, and potassium salts of amino ethanol phosphate (AEP)—which helps to support healthy cell membrane function.

The common diabetes drug metformin is also believed to kill senescent cells. In studies, metformin has been shown to alleviate a range of age-related disorders in animals and humans, including metabolic dysfunction, cardiovascular disease, cancer development, and cognitive dysfunction.[8] In elderly humans, it has been shown to increase lifespan by an additional five years.[9] Studies on mice show that these effects stem from reduced cellular senescence and fewer free radicals.[10]

Another exciting zombie killer is rapamycin. This drug inhibits a growth pathway called mammalian target of rapamycin (mTOR), which is responsible for regulating critical cellular functions such as cell growth, cell death, cell proliferation, and autophagy. Inhibiting mTOR appears to prevent the growth of senescent cells. In mice, rapamycin increases life-span, improves immune response, delays tissue loss, alleviates frailty with age, and decreases risk of heart failure, cancer, and cognitive impairment.[11] Not bad. Some doctors have quietly been using it as an anti-senescence drug since 2015.

I'm currently planning to experiment with taking rapamycin intermittently. It is not without risk—as we've discussed, any time you take a drug that accelerates cell turnover, there is a risk of accelerating cancer cell growth. In a couple more years, we'll know more about the risk-reward ratio, and it will be a lot more affordable. I'm always happy to be a guinea pig, and cutting-edge anti-aging people are using it today. However, unless you're in desperate straits, this is one hack you might want to hold off on for a little while as new research comes in.

And besides, there are other more affordable and more accessible natural compounds for fighting zombie cells. My favorite is fisetin, a polyphenol found in seaweed and strawberries. One study showed that high doses of fisetin could kill up to 50 percent of senescent cells in a particular organ.[12] While research on how to use fisetin to most effectively destroy zombie cells isn't complete, research indicates that it is a cognitive enhancer.[13] This is likely thanks to its direct antioxidant activity and ability to increase levels of other antioxidants in your cells. More antioxidants equals less oxidative stress and more energy throughout the body, including your brain!

It's not uncommon for researchers to discover that traditional herbs and plant compounds that have been used for thousands of years have anti-aging properties. A prime example is the Japanese herb ashitaba, which is available as a tea or powder and helps prevent zombie cells. It is traditionally used to treat high blood pressure, hay fever, gout, and digestive issues, but researchers recently discovered a compound in the plant called dimethoxychalcone (DMC—no relation to the famous rappers), which slows senescence. In worms and fruit flies, DMC increases life-span by 20 percent.[14] We don't know yet if it will have the same impact on humans, but it may be worth trying this tea to help with one of the Seven Pillars of Aging. I do.

Finally, there's piperlongumine (PPL), a pepper root extract that's commonly used in Ayurvedic medicine. It looks promising for reducing senescence, but this knowledge is so new that researchers don't yet understand the mechanism of action.[15] PPL may also have cancer-fighting properties, too, although research hasn't yet confirmed that benefit.[16] It is likely safe to use, but taking it all the time or taking high doses could put a load on the liver. If you do decide to use PPL consistently, your body might have a reduced capacity for detoxing[17]—so it's a good idea to take it for a limited period (one to two months) in conjunction with a liver-supporting supplement like glutathione.

It boils down to this: If you don't want to die, you must make sure your cells *do* die when they're supposed to and stay alive when they're supposed to.

PILLAR 4—CELLULAR STRAITJACKETS

The space in between your cells contains a network of proteins called the extracellular matrix, which protects your tissues from stress, trauma, and even gravity while allowing them to do their jobs. Visualize a perfect wobbling bowl of Jell-O. Without the matrix, you'd just have weird red liquid. Now imagine that same bowl of Jell-O, but so hardened that it won't wobble and you can't even spoon it. That's what anti-aging scientists call extracellular matrix stiffening.

Not only does the matrix literally hold your cells together, but it also gives your tissues their elasticity. This is incredibly important, especially when it comes to certain tissues such as those that make up the arteries. When these tissues lose their elasticity, they become stiff, and your body has to work harder to push blood throughout your circulatory system. This can of course lead to high blood pressure and heart disease.

So why does the matrix become so stiff? When sugar in your blood circulates throughout your system, it permanently binds with proteins, creating inflammatory advanced glycation end products, or AGEs. Glycation is the process of sugar bonding to protein. AGEs are aptly named, as these end products accelerate the aging process and create oxidative stress in the body.[18]

Think about it this way. When you eat something that contains sugar, glucose molecules travel through your body and look for proteins to bind with. Once stuck together, the glucose actually browns the proteins. This is the exact same chemical reaction that takes place when you brown onions in a pan and the sugar and onions become caramelized. When you have high blood sugar, it is at least partially because you made decisions that literally caramelized your insides. Yum. Not really.

There are multiple classes of AGEs. The most abundant in collagen is called glucosepane, and it contributes to diseases of aging from diabetes to vascular dysfunction. Thankfully, researchers are beginning to look for ways to break down AGEs and prevent them from stiffening the extracellular matrix. In 2018, the journal *Diabetes* reported that scientists had identified four enzymes that are able to break glucosepane cross-links.[19] They are still looking at the exact mechanism of action and whether or not the process of degrading AGEs creates other harmful metabolites, but this is a very promising area of research if you have type 2 diabetes or heart disease or just want to avoid this pillar of aging.

Even if, as I predict, glucosepane-degrading enzymes do prove to be safe and effective, it's better to just avoid extracellular matrix stiffening in the first place. To do that, you must reduce your blood sugar levels, particularly the spike in blood sugar you experience after meals. A study looking at glucosepane levels showed that this harmful AGE pretty much universally increases with age. In a nondiabetic control group, continuous high blood sugar more than doubled levels of this aging substance. Reducing blood sugar is not optional if you want to become Super Human. Fortunately, it's not as hard as you might think. You'll learn more about how to do this in chapter 3.

Chronic inflammation of any type is also associated with an increase in cross-linked proteins. This makes sense, since you already know that high blood sugar causes inflammation and high blood sugar causes these cross-links. In addition to managing your blood sugar levels, you should avoid eating foods that make you inflamed. When you are sensitive to a certain food, your body initiates an immune response that triggers inflammation. If this happens consistently, you end up with chronic inflammation and excess AGEs. There are good at-home tests that can help you pinpoint which foods you are sensitive to. I recommend Viome, which you'll read more about later, and EverlyWell.

(Disclosure: While I use both services, I'm an investor in and advisor to Viome, and EverlyWell has advertised on *Bulletproof Radio*.)

PILLAR 5—EXTRACELLULAR JUNK

As you age, waste products called extracellular aggregates build up both inside and outside your cells. Of the waste products that accumulate outside your cells, the main culprits are dysfunctional, misshapen proteins usually called amyloids. When amyloids start to accumulate, they stick together and form plaques that cause aging and disease by "gumming up the works" and getting in the way of healthy cellular interaction.

You can think of amyloids like the gunk clogging a sink. When you're young, you won't notice the impact—a single hair slips easily down a drain. But eventually, as more and more gunk accumulates in the pipes, water dissipates more and more slowly. It's that gradual process that slowly wears you down as you age.

You've probably heard that patients with Alzheimer's disease have a type of plaque (in this case called beta-amyloids, a type of protein aggregate) in their brains. But long before you develop Alzheimer's, these same plaques can impair cognitive function. In the case of type 2 diabetes, one type of protein aggregate called islet amyloid inhibits insulin secretion. Protein aggregates also cause stiffening in the heart. This is called senile cardiac amyloidosis and is a major cause of heart failure.

So what causes proteins to stick together in the first place? The problem with amyloids is that they build up in different tissues for different reasons, and we don't know all the reasons yet. We do know that autoimmunity, when the immune system attacks its own healthy cells, makes it worse, and at least 30 percent of people have some form of autoimmune disease. And recent research on mice links *low* insulin levels to the formation of amyloids in your brain.[20] This is one reason you don't want to be on an unending low-carb diet that keeps you in ketosis without pause. You'll live longer if you sometimes eat low carbs, sometimes eat moderate carbs, and always avoid sugar and bad fats. Low insulin is worse than high insulin in this case, but neither will keep you running at your peak.

Even if you don't have full-blown autoimmunity, inflammation stemming from food sensitivities or even unending emotional stress can lead to amyloid buildup (in addition to AGEs). It appears that amyloids form during long periods of chronic inflammation from any cause. The smart strategy is to reduce your inflammation levels by avoiding foods you are sensitive to and learning how to chill out. If you're eating food that's not compatible with your biology, you're going to end up inflamed, and that will age you in multiple ways. Same deal if you spend a lot of time in a state of stress.

The good news is there are simple strategies you can use to partially break down or reduce the formation of these proteins that age you prematurely. One of the best things you can do is to boost autophagy, your body's recycling program, by consuming more of the foods you will read about in the next chapter. This will help break down these proteins so they don't end up forming harmful plaques. So will fasting.

Gordon Lithgow, PhD, a professor at the Buck Institute for Research on Aging, has also found that vitamin D helps prevent proteins from losing their shape and sticking together. With vitamin D deficiency so widespread,[21] this raises the question of whether Alzheimer's rates are increasing in part because people do not have enough vitamin D to slow amyloid plaque formation.

There is also a clear connection between toxic heavy metals and amyloids. A study from the Society for Neuroscience found that excess copper prevented the body from clearing protein aggregates on its own.[22] You need copper for many functions in the body, but too much of it is toxic. Medical research shows that the blood vessels and brains of patients with Alzheimer's disease contain excess copper. Cadmium, another heavy metal, increases the formation of protein aggregates in the brain and appears in greater amounts in brain tissues of patients with Alzheimer's disease than in healthy brains.[23] You'll learn how to avoid and detox from these metals and others later in this book.

In his lab, Lithgow has demonstrated that chelators, small molecules that bind with heavy metals and help you detox, protect mice from developing protein aggregates. You won't be surprised to hear that chelating from heavy metals has been a priority of mine for years. You'll read all about how to do this later. Heavy metal exposure has been on the rise for decades, and no matter where you live or how clean you eat, chances are that you still have higher than ideal levels of metals like lead and mercury. Approximately 6 million pounds of mercury is released into the environment each year, and lead, arsenic, and cadmium are present in detectable levels in our air, water, food, medicine, and industrial products. Even organic kale is high in one heavy metal.

In addition to contributing to the buildup of amyloids, heavy metals also cause mitochondrial dysfunction.[24] A small amount of exposure to lead, mercury, nickel, uranium, arsenic, or cadmium for a short amount of time can impair mitochondrial energy production and increase mitochondrial death.[25] Even if you don't realize it, the heavy metals already in your body are likely aging you right now. You'll learn about how to detox them later.

PILLAR 6—JUNK BUILDUP INSIDE CELLS

Okay, so waste products can build up outside of your cells, but the good news is that nearly all the cells in your body have their own built-in waste disposal system called a lysosome. Your lysosomes incinerate unwanted materials of all kinds, keeping your cells free of junk and able to function optimally.

You knew there was a *but* coming, right? When the lysosome can't break down certain materials to incinerate them, the waste products end up just sitting there, clogging up the cell until it can no longer function. The name for this is intracellular aggregation. If this happens to too many of your cells, you end up with Pillar 1—loss of cells and tissue atrophy.

There are two reasons this might happen. The first is if the lysosome itself is damaged and can't function properly. Lysosomes rely on over sixty types of enzymes to break down waste products, and mutations in the genes for these enzymes can prevent the lysosome from doing its job. These organelles can also be damaged by an excess of reactive oxygen species—free radicals—which happens when your mitochondria aren't working efficiently.

But the more likely reason your cells fill up with junk is that you eat too many foods that your lysosomes are incapable of incinerating even if they are functioning perfectly. These are advanced glycation end products (AGEs) that you eat rather than the ones that are made by sugar inside your body. Remember when I said that when sugar and proteins link up inside your body, it is the same as caramelizing onions? Yeah, it also happens when you eat caramelized protein: aka charred meat (from grilling over an open flame, broiling, or cooking protein with sugars). The AGEs you consume get stuck inside your cells, and your lysosomes can't clear them out.

Over time, these materials build up, making more and more of your cells dysfunctional, and this affects your ability to control blood sugar levels[26] and increases your risk of cancer[27] and heart disease.[28] When it happens to neurons, it can contribute to Alzheimer's.[29]

Fried, blackened, and charred meat all contain tons of AGEs that can overload your cellular waste system and leave your cells literally full of garbage. And this dramatically raises your risk of developing one or more of the Four Killers. A 2019 study published in *BMJ* looked at the dietary habits of over one hundred thousand women between the ages of fifty to seventy-nine over the course of several years. After taking into account potentially influential factors such as lifestyle, overall diet quality, education level, and income, the researchers found that regularly eating fried foods (which also contain AGEs, since frying produces a similar chemical process as charring meat) was associated with a heightened risk of death from any cause and, specifically, heart-related death. Those who ate just one or more servings of fried food a day had an 8 percent higher risk of death from heart disease than those who did not eat fried food. One or more servings of fried chicken a day specifically was linked to a 13 percent higher risk of death from any cause and a 12 percent higher risk of heart-related death than someone who ate no fried food.[30]

This one hurts, I understand. When I was in my twenties, I was the master of the grill. I loved charring meat over an open flame, but now

I love my clean, highly efficient cells even more. It's worth ordering grass-fed steak with no char.

PILLAR 7—TELOMERE SHORTENING

Take a moment to picture the plastic tips on the ends of shoelaces that protect them from fraying. Your telomeres serve a very similar function—they are the endcaps of your DNA that protect your chromosomes from fraying with wear and tear (aka age). An enzyme called telomerase is responsible for maintaining telomeres, but these caps naturally deteriorate over time because each time a cell copies itself, the telomeres shorten. As you age, they get shorter and shorter until they can no longer protect the cell. The cell then either stops growing or submits to apoptosis. In fact, there is a term for the number of times a cell can divide before it is no longer protected by telomeres and dies—it's called the Hayflick limit.[31]

Shortened telomeres are linked to a weakened immune system and chronic and degenerative diseases like heart disease and heart failure,[32] cancer,[33] diabetes,[34] and osteoporosis.[35] The rate at which your telomeres shorten plays a huge role in determining the rate at which you age. Scientists view telomere length as a reliable marker of your biological age (as opposed to your chronological age). People with shorter than average telomere length for their age have a higher risk for serious disease and early death[36] than their peers with longer telomeres. In one study, people over the age of sixty with shorter than average telomeres had three times the risk of dying from heart disease and eight times the risk of dying from an infectious disease[37] as someone with average-sized telomeres for their age.

It's clearly critical to keep your telomeres long. There are some studies showing ways to lengthen telomeres, but not enough evidence yet to say that we know for sure how to do it in every case. But we do know some things about what make telomeres shorter and how to protect them from shortening. Interestingly, there seems to be a direct connection between telomere shortening and stress. In one study, women with the highest levels of perceived stress had telomeres that were shorter by the equivalent of one full decade than women who said they experienced less stress.[38] This is an important finding because it offers evidence that how you experience psychological stress has as much of a physiological impact as environmental stress. And this makes sense, since both psychological and physiological stresses are associated with increased oxidative stress in the body.

Exercise is another important way of preventing early telomere shortening. Researchers in Germany looked at telomere length in four groups of people: those who were young and sedentary, those who were young and active, those who were middle-aged and sedentary, and those who were middle-aged and active. There wasn't much of a difference between the two groups of young people, but when the participants were middle-aged, the change in telomere lengths was striking. The sedentary middle-aged folks had telomeres that were 40 percent shorter than the young people, while the active middle-aged folks had telomeres that were only 10 percent shorter than the young people. In other words, the active group reduced their telomere shortening by 75 percent.[39] Exercise significantly reduces perceived stress levels and inflammation,[40] which may help to explain these results.

There are two promising lines of research about lengthening telomeres. One is a synthetic peptide called Epitalon that is modeled after a peptide your pineal gland produces (epithalamin). The research on Epitalon goes back to 2003, but no one has commercialized it. When researchers injected Epitalon into mice, it was shown to increase their life-span by up to 13.3 percent by activating telomerase[41] while turning on apoptosis and slowing down tumor growth.[42]

Someone with identical biology to me (ahem) has been injecting Epitalon for ten days every few months for the past several years despite the fact that it is not yet approved for human use and may never be, even though it seems to work. In fact, anti-aging substances like Epitalon often exist in a strange limbo. The pharmaceutical companies don't develop them because they're not patentable, which means they won't pay for the huge studies the FDA requires before approving them. The result is that you can find Epitalon affordably online, but it's hard to know that you're getting it from a reputable source. To me, the risk-reward ratio is worth it, but this may not be the case for you.

Another supplement called TA-65, the name brand of cycloastrag-enol, also activates telomerase.[43] It is incredibly concentrated extract of an Ayurvedic herb called astragalus. By law, the makers of TA-65 can't call it an "anti-aging" drug because it hasn't been proved to extend life-span. But studies on this molecule show that in humans, it improves biological markers associated with health span through the lengthening of telomeres and rescuing of old cells. The downside here is that it is quite expensive. If you've experienced a lot of stress and/ or feel that you are aging more quickly than you'd like and it's in your budget, this might be worth considering. There are generic versions available, too.

Until we know more about how to maintain telomere length, avoiding excessive environmental stress and taking measures to reduce your psychological stress is a good start, along with getting good quality sleep to recover from stress that is truly unavoidable.

You've probably noticed that these simple interventions—good food, the right environment, moderate exercise, stress control, and quality sleep—are the best and most effective ways of avoiding all Four Killers and even slowing down or reversing many of the Seven Pillars of Aging. And you're right! The vast majority of the hits to your mitochondria that cause aging come from your food, your environment, and your lack of quality sleep. So, before we move on to aging backward, let's take a closer look at the most important ways to avoid dying. After all, what's the point of being a Super Human if you're dead?

Bottom Line

Want to not die? Do these things right now:

• Kill off death-resistant cells with natural or pharmaceutical compounds such as AEP, fisetin, and piperlongumine.

• Consider getting anti-aging drugs like rapamycin or metformin from your doctor.

• Stop eating fried, grilled, or charred meat. It's just not worth it if you want to live a long, high quality life.

• Manage stress—meditate, practice yoga, get good quality sleep, and/or delegate tasks that are draining you. This is not indulgent or selfish—it will literally help you live a longer and fuller life.

• Consider supplementing with vitamin D to help your body avoid forming dangerous misshapen proteins.

• Do what it takes to find out which foods are not compatible with your biology, either through an elimination diet or a food sensitivity panel, and stop eating those foods.

FOOD IS AN ANTI-AGING DRUG

By the time it became clear that inflammation made me feel like crap and was aging me rapidly, I had conducted enough semi-successful experiments on myself to know that of all the things I could control, food had the biggest impact on how I felt, how I performed, how inflamed I was, and therefore how quickly my body aged. Armed with this experience and the lifetimes of knowledge distilled from medical reports, biochemistry, and experts at SVHI, I set out to determine once and for all which foods and compounds supercharged my mitochondria and reduced inflammation and which led to inflammation, dysfunctional mitochondria, and rapid aging. Fortunately, most of the good stuff also tasted good!

Years later, I wrote *Game Changers*, based on a survey of almost five hundred people who had done big things in the world—I wanted to figure out what made them tick, what qualities these superstars had in common. The results showed that high-performing people know that getting their food right is the number one human upgrade, even though different people found that different foods worked best for their individual biology. Nutrition is essential not only for Super Human biology but also for Super Human success.

GRAINS, GLUTEN, GLUCOSE, AND GLYPHOSATE (OH MY)

In my mid-twenties, I figured out how to lose fifty pounds of fat, decrease inflammation, gain energy, and gain positive changes to my personality using multiple versions of a low-carbohydrate, high-protein diet. I was happier and less angry, and had more friends and more energy. It was clear that something in my diet had caused these drastic changes. As I experimented with different types of carbs, I realized that for me, gluten was the number one culprit. Even though I do not have celiac disease, a condition that makes the small intestine hypersensitive to gluten, my body did not tolerate gluten well, and responded with chronic inflammation and changes to my personality that were far from positive.

Chances are you've already heard about the damaging effects of gluten, along with strident, shrill misinformation about how only people with celiac disease should avoid it. The sad truth is there is plenty of research to show that eating wheat—not just gluten, the protein found in wheat—is aging for the rest of us, too. Wheat causes inflammation and gastrointestinal distress and contributes to autoimmune disease and a host of other issues by stimulating an over-release of zonulin, a protein that controls the permeability of the tight junctions between the cells lining your gut. It does that whether or not you tell yourself that you tolerate wheat just fine.

With excess zonulin, the gaps between your intestinal cells open, allowing bacteria, undigested food, and bacterial toxins to flood into your bloodstream. Those toxins, called lipopolysaccharides, or LPSs, lead to inflammation throughout your body. They make you old,[1] and as you get older, the accumulation of hits from LPSs impacts your health more and more.[2] They do this no matter what you think about gluten.

Gluten also reduces blood flow to the brain, interferes with thyroid function,[3] and depletes your vitamin D stores.[4] As you read earlier, vitamin D deficiency can cause proteins to lose their shape and clump together, forming dangerous and aging plaque deposits.

If you've been following the latest news about gluten, you're probably confused. On one hand, the Big Food industry says to eat it, but if you're listening to the frontline anti-aging doctors on *Bulletproof Radio*, you hear clear advice to avoid gluten. You may have even switched to other grains besides wheat to avoid gluten. Unfortunately, most grains contain plant compounds designed to weaken animals like us who eat them. They also commonly contain storage toxins and field toxins from mold that grows on crops, and grains are commonly sprayed with glyphosate, the main ingredient in the herbicide Roundup.

In May 2015, the World Health Organization (WHO) classified glyphosate as "probably carcinogenic to humans" based on animal studies showing that glyphosate caused tumor growth and higher incidents of cancer. The WHO investigation also found that glyphosate is probably genotoxic (meaning it causes mutations in DNA) and increases oxidative stress, which triggers inflammation and speeds up aging. Glyphosate also mimics estrogen, which might explain why it causes human breast cancer cells to grow in vitro.[5] Roundup itself is directly toxic to mitochondria[6] and even more toxic to human placental cells than glyphosate[7] alone.

Even more worrisome, the *gly-* in glyphosate stands for glycine, an amino acid prevalent in collagen, the protein in your skin's connective tissue. Glyphosate is actually a glycine molecule attached to a methylphosphonyl group (which happens to be a precursor to chemical weapons). This means that when you consume glyphosate it can be incorporated into your collagen matrix just like glycine. In 2017, the Boston University School of Public Health released research showing that glyphosate substituting for glycine disrupts multiple proteins necessary for kidney health and may contribute to kidney disease.[8] Plus, your skin is made of collagen. Extra wrinkles won't necessarily keep you from living longer . . . but it's always nice to look as young as you feel.

Before we spread another 18.9 billion pounds of glyphosate on our planet, we simply must conduct more research on how glyphosate contributes to other diseases when the body uses it as a substitute for glycine. For now, suffice it to say that if you want to avoid the painful, slow decline we now associate with aging, avoid glyphosate, which means avoid grains (at least in the United States). That's not as easy as it may seem. Not only are the vast majority of conventionally grown grains sprayed with Roundup, but so is much of our conventionally grown produce and the grains that are fed to conventionally raised animals. This means glyphosate is hiding in most products containing corn and other grains, industrial feedlot meat, and animal products like nonorganic milk, yogurt, cheese, and so on.

Many parents were rightfully horrified when a 2018 report showed small but meaningful amounts of glyphosate in name brand breakfast cereals and other products marketed as healthy choices for families. I am equally horrified when I see advertisements for bone broth made from nonorganic industrial chickens. While bone broth is a great source of collagen, when it is made from the bones of conventionally raised chickens, it is a glyphosate land mine.

The good news is that the executives running Big Food companies will change how they make food when you demand it. After all, they have kids and don't want to get old, just like the rest of us. This is simply about getting the science into the hands of decision-makers and getting them to believe it. Having had the opportunity to sit down with the heads of many of the largest food companies, I can attest that they feel a moral and personal obligation to feed you the healthiest food that you will actually eat at the lowest possible cost. They are good people who want to do the right thing. That is real. These are not evil people (except the people still making glyphosate . . . there must be a special place in hell for them). It's just that we haven't shown Big Food companies that we will actually pay a tiny amount more to get food that keeps us young. It's okay. They're coming over to our side as the data becomes clearer.

Glyphosate is just one reason that where you get your food really matters. After years of thinking about it, I decided to make the difficult move to an organic farm where my family can grow our own produce and even raise our own animals (and trade with neighbors who raise animals we don't). But even before I made the move, my health improved tremendously when I simply eliminated grains and switched to organic, grass-fed animal products from the grocery store and farmers' markets.

Despite these changes, I still had to learn how to control my blood sugar. More to the point, given how aging high blood sugar is, I learned how to kick its ass. On so many of the diets I'd tried in the past, I ate a low-fat, low-calorie, high-carb breakfast. My body secreted insulin to transport sugar to my cells so they could create energy. This caused a spike in blood sugar, then a quick drop until my base instincts yelled at me to eat something quick to get more energy. Sound familiar? Sugar cravings are how our biology evolved to keep us from starving to death, but they certainly weren't helping me live longer! Even short spikes in blood sugar cause damage to the inside of your arteries, contributing to cardiovascular disease.

Another common scenario was that I'd unknowingly eat something that contained toxins, requiring my liver to use extra energy to filter toxins out. This of course led to more sugar cravings as my liver struggled to make enough energy to oxidize the toxins. My entire life was ruled by sugar cravings! It had been for as long as I could remember. And when I gave in and ate the darn sugar (or refined carbs), of course it made things worse: More blood sugar means bigger energy crashes, more oxidative stress,[9] and the constant formation of AGEs when all that sugar links up with protein in tissues. You already know that sugar ages you, but you may not know how to stop eating it or about the deadly combination of too much sugar and too much protein . . .

THE VEGAN TRAP

Then I read *The China Study* by T. Colin Campbell and Thomas M. Campbell II, one of the first popular books that made the connection between eating animal products and many common diseases, including the Four Killers. According to an uncritical read of the book, the best way to avoid dying is to avoid animal products completely. Given that not dying is the first step to anti-aging, and not having done all the research, I decided to avoid animal foods.

So I turned to a raw vegan diet, and I went all in. I bought sprouting trays and the world's best blender and spent my days eating bowl after bowl of salad and entire blenders full of green smoothies trying to consume an adequate number of calories. It worked . . . for a little while. I got down to about 185 pounds—too low for a six-four guy—and I felt a burst of energy that also made me feel flighty and ungrounded. I convinced myself that the increase in pain and stiffness that came with this was just my body "detoxing." But my friends said I looked gaunt, and pretty soon I wasn't feeling so great. My teeth got sensitive and even started to break, and I felt cold all the time. It was pretty clear that I was suffering from malnutrition, despite knowing a ton about nutrition and spending two hours a day preparing food.

Later I learned about what I call the "vegan trap." When you switch from a diet containing animal fats to the mostly omega-6 polyunsaturated fats found in plants, you set yourself up for failure. Vegetable oils reduce your thyroid function by preventing thyroid hormone from binding to receptors.[10] At first your thyroid hormones will temporarily increase to compensate for the lower energy, and you feel good. That's what led to my ungrounded energy and initial weight loss. But if you continue to give your body the wrong building blocks, your health will suffer. Because your cells don't have the right building blocks to make energy efficiently, your metabolism eventually slows down. That slow metabolism doesn't just make you gain weight more easily; it slows down your brain, your energy, and everything you do.

For about six weeks as a vegan, I felt great and grew convinced that my diet was the answer to all my problems. I had tons of energy and no idea it was the energy of a stressed animal that is starving and needs one last boost to catch its prey. Already convinced that being a vegan gave me more energy, I logically decided to "lean in" when I started to feel the ill effects. That is why it's a trap—once you're convinced that you feel good on a vegan diet (because you actually did for a short while), you don't think to look at your diet when your energy or your health begin to suffer.

Thankfully, it took me only about six months to realize what was going on, do more research, and decide to add meat back into my diet. By then I had learned about the dangers of consuming AGEs when eating overcooked meat, so for a brief period of time I became a raw omnivore. In addition to occasionally eating sushi, I marinated thin strips of steak in apple cider vinegar to kill harmful bacteria and added it to my salads. With that plus some raw egg yolks and raw butter, I started to feel better right away.

When I reread *The China Study*, I realized it had some serious flaws. For example, the researchers conclude that all animal protein causes cancer simply because rats that were exposed to large amounts of casein (a dairy protein, one of thousands of animal proteins that each do different things) had a higher chance of developing liver cancer than rats that did not consume casein. But the study didn't account for the type of animal product or the type of animal; nor did it consider what that animal ate or how the meat was stored or cooked. These factors truly determine whether or not an animal product is aging. So does the amount of meat you eat. If you want to live a long time, you want to avoid eating too much meat and avoid eating all low quality meat.

My time as a raw vegan was not fun, but I am grateful for *The China Study*. Had I not cut out animal protein from my diet, I wouldn't have become familiar with the research showing that most of us—including me before I went vegan—eat far too much protein in general. Eating a pound of steak or chicken every day has a different impact from eating a few ounces, which in turn has a different impact from eating none at all.

After I started eating meat again, I wondered why my inflammation levels had decreased when I cut out animal products. It turns out excess protein—especially from animals—causes inflammation. Most animal protein contains specific amino acids such as methionine, which causes inflammation and aging when eaten in excess. (Except for collagen protein, which has far less methionine.) In pharmaceutical studies, this is called an inverted U-shaped response curve. It means there is a "Goldilocks zone" for dosing a substance, and either too little or too much does not work.

This is no small consideration. When you eat a diet high in animal protein, you can expect a 75 percent increased risk of dying from all causes over eighteen years, a 400 percent increased risk of dying of cancer, and a 500 percent increased risk of diabetes compared to someone who restricts his or her animal protein.[11] Totally not Super Human. Another set of studies found that restricting protein can help increase maximum life-span by 20 percent, probably because less protein means less methionine.[12]

The type of protein you eat is as important to consider as how much protein you eat. If the protein in question is charred or deep-fried, there is no good amount to eat. Same goes if it's from industrially-raised animals treated with antibiotics. But if the protein is from gently cooked grass-fed animals, wild fish, or plants (hemp is best), then there is a simple formula for the correct daily allowance: about 0.5 grams per pound of body weight for lean people; and about 0.6 grams per pound for athletes, older people (the risks associated with overconsumption of protein decrease after age sixty-five), and pregnant women.

If you're obese like I was, sorry, but all that extra fat you're carrying around doesn't require protein, so subtract it from your body weight before figuring out how much protein to eat. For instance, when I weighed 300 pounds, let's assume I was carrying an extra 100 pounds of fat. Take my weight (300), subtract my fat (100), and you end up with 200 pounds, so I should have aimed to eat 100 grams of protein (0.5 × 200 pounds). If you're relatively heavy and have no idea of your body fat percentage or are just bad at math, assume you're about 30 percent fat. So you'd eat 0.35 grams per pound of body weight.

Collagen protein is a special case. Given that it lacks the most aging amino acids and has all sorts of benefits for connective tissue, you can add another 20 or more grams of grass-fed collagen on top of your protein intake or use it as part of that number. Some days up to 50 percent of my protein comes from Bulletproof collagen.

Eating less protein will not give you less energy. Contrary to everything you've heard from most popular diets (even keto), protein is actually a terrible last-ditch fuel source for humans, worse than fat or carbohydrates. The process of turning amino acids from protein into energy creates a lot more waste than fat or carbs, and excess protein ferments in the gut and produces ammonia and nitrogen. This puts a huge load on the kidneys and liver. Instead of getting energy from protein, you want to consume just enough protein as building blocks to repair your tissues and maintain muscle mass, and then get energy from fat, fiber, and a few carbs, instead.

When you get this right, your cells can rebuild themselves with clean animal fats and protein (notice, you're an animal, too), and your gut bacteria will actually transform fiber from vegetables into fatty acids, an ideal fuel source for your mitochondria. Add in excess protein, antibiotic-contaminated meat, and/or sugar, and your gut bacteria just won't do the same thing.

Restricting protein intake also helps boost autophagy, your all-important cellular recycling program. By occasionally limiting how much protein you eat (you can still have a nice steak every once in a while), you force your cells to find every possible way to recycle proteins. In their search, they excrete waste products hiding in your cells, slowing down energy production. Temporary protein deficiency is a type of hormetic (beneficial) stress. In response to protein restriction, your body looks for other sources of energy. It is the equivalent of burning your trash to stay warm.

The same thing happens when you use intermittent fasting (simply eating all of your food within a shortened period of the day, usually between six to eight hours) as a type of hormetic stress. Intermittent fasting is incredibly useful in aiding fat loss, preventing cancer, building muscle, and increasing resilience. Done correctly, it's one of the most painless high-impact ways to live longer.

Until recently, we did not fully understand why fasting was so beneficial. Then in 2019, scientists at the Okinawa Institute of Science and Technology discovered that just fifty-eight hours of fasting dramatically increases levels of forty-four different metabolites, including thirty that were previously unrecognized.[13] Among other beneficial functions, these metabolites—substances formed during chemical processes—boost antioxidant levels in the body. And as we know, antioxidants are important for fighting off aging free radicals. All of these benefits can be explained by the fact that fasting dramatically boosts autophagy,[14] keeping your cells young and healthy.

Fasting has profound effects, even at less than fifty-eight hours. Alternative day fasting, a form of intermittent fasting in which you eat every other day, helps prevent chronic disease and reduce triglyceride and low-density lipoprotein (LDL) cholesterol levels in as little as eight weeks.[15] Intermittent fasting also increases your brain's ability to grow and evolve by boosting neuronal plasticity (the brain's ability to change throughout your life) and neurogenesis (the birth of new neurons).[16] This can help ward off Alzheimer's and cognitive decline.

As you might expect, when I started experimenting with intermittent fasting ten years ago, I was often left feeling cranky and cold around lunchtime, before my eating window opened. This is because I had not yet developed the metabolic flexibility from teaching my body to efficiently burn carbohydrates *or* fat. Today I can effortlessly fast for twenty-four hours because my metabolism is younger and my blood sugar levels have stabilized. Thankfully, there are now well-understood ways to make intermittent fasting painless, which you'll read about later.

A BIG FAT LEAP OF FAITH

So, when it comes to aging, grains are bad, sugar is bad, fried stuff is bad, and too much or too little protein is bad. What about fat? Can you eat too much of it? Sure. But we need fats for reproductive health, temperature regulation, brain function, and shock absorption. Fat helps build the outer lining of your cells, which protects them from damaging substances. It also makes up the bile acids you need to digest foods, and vitamins A, E, D, and K are fat soluble, meaning your body needs fat to absorb them. Additionally, several important hormones, including leptin, which helps you feel satiated, are made from saturated fat and cholesterol. Fat is also the basis for the lining of your nerves, called myelin, which allows electricity to flow efficiently between nerves and is essential for avoiding degenerative diseases like multiple sclerosis.

Saturated fat in particular is so important that your body converts carbs to palmitate, a type of saturated fat, in a process called de novo lipogenesis. Without this ability, you'd die. That's how critical saturated fat is. Your body then converts palmitate into other saturated and monounsaturated fats necessary for cell membranes, but it can't make enough polyunsaturated omega-6 and omega-3 fats. That's why you have to eat them. Yet the myth that eating fat and cholesterol will make you fat and give you heart disease still somehow persists. You read earlier that it's your gut bacteria and not dietary cholesterol that creates plaques that build up in arteries. The evidence is in, and the fats you eat that contain cholesterol are not the enemy, as we've been told.

When you eat enough of the right fats without excess carbs or protein, your body learns to efficiently burn fat for fuel. If you eat excess carbs or protein, your body burns those first. Normally, your body converts carbohydrates to make glucose, which your mitochondria use to produce energy. When you run out of carbohydrates, you start converting fat to glycerol for energy. The liver produces ketones as a by-product of this fat metabolism, and your mitochondria burn those ketones instead of glucose in a more efficient form of energy production. Ketosis is a state your body enters when you have a lot of ketones in your blood and are burning additional fat . . . or when you eat a special type of saturated fat that converts to ketones in your body. More on that later.

One last time: Your body requires fats for you to perform your best and live as long as possible. You just have to know which fats serve what purpose. Some fats you eat are building blocks for your body, and some are better used as fuel. Getting the mix right matters. But have you ever heard nutrition "experts" say exactly which of the many saturated (or other) fats to avoid? The typical buckets you hear ("plant based," "animal fats," "saturated," "polyunsaturated") are not very specific. Is it possible that the heated industrial polyunsaturated fat in French fries has a different effect on your biology than avocado oil, or that the fat in industrially-raised animals is different from the fat in an egg yolk or pastured beef? You bet it is.

Researchers in Australia have measured how different cells elegantly use each type of fat you eat. You can make sure your brain has the type of fuel it runs best on and that your body fat doesn't create extra inflammation and make you old. Eating the right fats could add productive years to your life, which is why it's worth a page or two of your time to dig a little deeper into details of how your body uses fats.

Scientists describe cell membranes as "the margin between life and death for individual cells."[17] These membranes are made of tiny droplets of fat. About 5 percent of your genes contain instructions telling your cells how to make the thousands of types of fat your body needs to survive. We now know so much about what each different type of fat does that French researchers have proposed the notion that "saturated fats should no longer be considered as a single group in terms of structure, metabolism, and functions."[18] In other words, we have grouped together a very diverse array of fats under one reductive and often misleading label. When your doctor tells you to eat less saturated fat, your response should be "Which one(s) do you mean?"

I've had the opportunity to interview lots of fat experts (or experts on fat), and most of us use an analogy from nutritionist and early trans fat researcher Mary Enig, PhD, who popularized two basic ways of thinking about the fat you eat. The first is to look at how long a fat molecule is. There are short-chain, medium-chain, and long-chain fats. As a general rule, the shorter the saturated fat, the more anti-inflammatory it is. For instance, butyric acid, which is anti-inflammatory, has only six molecules, while other types of fat may have twenty or more.

Some fats are easy to damage no matter how long they are. So the second way to understand your fat is to assess its stability. Oxygen drives very strong chemical reactions that damage fats through oxidation. Oxidized (damaged) fats cause you to age more quickly by creating inflammation in the body and building less effective cell membranes. When your body has no choice but to incorporate oxidized fats into cell membranes, those cells create excess free radicals that make you an average human, not super.

Your cells use saturated fats, which are the most stable of the fats, to make about 45 percent of the cell membranes in the brain and liver, and about 35 percent in heart and muscle cells.[19] Yes, saturated fat is the dominant fat in your brain, so don't demonize it! Energy-producing cells will hold their level of saturated fat at about this level *no matter what type of fat you eat*. The only type of tissue that meaningfully changes its composition of saturated fat is adipose tissue— aka your muffin top. When you eat more saturated fats, the cells in adipose tissue will change their makeup to contain more saturated fat and less unstable fats without changing in size. This is fantastic, as stable fats make for fewer free radicals.

Think of saturated fat as the stable waxy bricks building the "walls" for your cells. The problem is that your cell membranes have to flex in order to make energy and receive chemical signals, and those nice stable saturated fat "bricks" don't bend. So while it's fine to go ahead and eat butter and other forms of saturated fat, it's also important to eat other types of fats. And those include the next most stable group of fats, monounsaturated fats. These fats—found in food sources like olive oil, avocados, and some nuts—are more flexible than saturated fats. You can think of them as the gel-like "mortar" that supports your saturated fat bricks in the cell wall. Your cell membranes are made up of about 20 percent monounsaturated fat.

Interestingly, brain cells have the most monounsaturated fat of any cells in the body, and they hold their level of monounsaturated fat constant no matter what types of fat you eat. Most other cells adjust their fat content slightly when you eat a lot of monounsaturated fats.

But without changing how much fat you have on your body, fat cells will happily dump other stored fats and replace them with monounsaturated fat. This means you can transform your stored body fat to have a higher percentage of stable fats. Eat your olive oil!

After you account for the saturated and monounsaturated fats in the membranes of energy-producing cells like muscle, you're left with about 35 percent of a combination of polyunsaturated omega-6 and omega-3 fats, as well as some conjugated linoleic acid (CLA), a type of fat produced by microbes in your gut. (CLA also happens to be found in grass-fed butter—more on this in a bit.) While omega-3 and omega-6 fats fall under the same category, they are not the same.

Omega-3s are anti-inflammatory and thus beneficial to your anti-aging efforts. The best omega-3 fats are found in food sources like cold-water fish (salmon, mackerel). You can also get omega-3s from walnuts and olive oil, but vegetable omega-3s are only 15 percent as effective as those found in fish.[20]

Unfortunately, omega-3 fats are far outnumbered by omega-6s in the standard Western diet—and omega-6 fats are highly inflammatory. Poultry, the most common protein in Western diets, is high in omega-6s. Most refined vegetable oils are also polyunsaturated omega-6s, and they are so unstable and inflammatory that eating excess canola, corn, cottonseed, peanut, safflower, soybean, sunflower, and all other vegetable oils is likely to contribute to cancer and metabolic problems. Oxidized omega-6 fats damage your DNA, inflame your heart tissues, raise your risk of several types of cancer, and don't support optimal brain metabolism.[21] Anything that increases inflammation decreases brain function.

When you cook with those fats, they are even more aging because they become oxidized so easily. Remember how aging oxidative stress is? Eating oxidized fats speeds this process way up. Additionally, trans fats are a category of omega-6 fats that are the most dangerous of all. Decades ago, when food manufacturers needed a shelf-stable fat for processed foods, they created hydrogenated omega-6s, or trans fats. These fats are linked to many health problems and cause obesity, and it took the food industry only forty years from the time they learned about this to begin phasing them out. When you ingest man-made trans fats, your body tries to use them to build cells, but cell membranes made of these trans fats cannot function properly. And without healthy membranes, you'll never make it to a hundred and eighty—or even a comfortable seventy-five.

Artificial trans fats also form when you use polyunsaturated fats for frying.[22] Fortunately, trans fats won't likely cause problems if you use the oil for frying only once, but restaurants often use the same oil over and over all day or all week, which creates oxidized oil *and* trans fats. So put down the French fries, no matter how lean you are. Seriously—you're better off having some rum or smoking a cigar. Super Humans don't eat fried food, even if it's crispy and delicious. You know what's not delicious? Eating from a tube later because you couldn't put down the chicken wings when you were younger.

Your body does need *some* omega-6s, but there are so many of them in a standard Western diet that you would have to work really hard to consume too few. Ideally, you should consume no more than four times as many omega-6s as omega-3s, but most people today eat an average of *twenty to fifty times* more omega-6s than omega-3s. This is a hugely underreported source of accelerated aging. Changing the balance of omega-3s to omega-6s you consume can give you a Super Human metabolism because your stored fat cells change *dramatically* when you eat omega-6 fats. No matter how much (or how little) body fat you have, anywhere from 7 percent to 55 percent of it is made of inflammatory omega-6 fat, depending solely on how much of each type of fat you eat.

If you are lean, you want to eat the same composition of fats that you want stored in your body. That means that whether you're on the high-fat Bulletproof Diet or a low-fat diet, stick to about 50 percent saturated, 25 percent monounsaturated, 15 to 20 percent undamaged (meaning not oxidized) omega-6, and 5 to 10 percent omega-3 fats, including EPA and DHA. If you are obese and have a good amount of excess body fat (like I used to have!), right now your body is probably storing too many unstable fats. To shift your fat composition, temporarily eat an even higher percentage of the type of fats you want in your body. Of the fat you eat, 50 to 70 percent should be saturated, 25 to 30 percent monounsaturated, and only 10 percent undamaged omega-3 and omega-6.

The challenging thing is that the most common blood tests doctors use to measure things like cholesterol and triglyceride levels do not offer an accurate picture of the type of fats in your brain, heart, or muscle cells, which is different than fat in your blood cells. So there is good reason to distrust the fat ratios found in the blood tests that most doctors rely on. Looking at inflammation markers in your blood work, such as C-reactive protein (CRP) and homocysteine, will give you a much more accurate sense of how you're aging.

When I started experimenting with eating more fat, I was nervous— it went against everything I'd been told about healthy eating. One of the biggest leaps I took was to begin eating more grass-fed butter. When I took a deep breath and stopped holding back on butter, magical things started to happen. My focus increased, I had more energy, and my blood panels showed that my levels of inflammation had decreased.

Like any good biohacker, I kept experimenting until I knew I had taken things too far. I heard that some Inuit populations survived on no carbohydrates at all, so I decided to subsist on a diet of almost entirely fat and animal protein and see what it would do for my health and performance. The result of that experiment was a host of new food allergies because the bacteria in my gut were literally starving and out of desperation began eating my own gut lining. Sadly, a diet of only steak and butter won't work for the long term. But it was delicious in the short term.

PIG'S EARS AND ENERGY FATS

By implementing everything I'd learned about nutrition, I was able to dramatically decelerate my aging. My knees were still a mess, but I weighed less and had more energy than ever before, and I managed to (barely) graduate from business school while working full time despite my cognitive dysfunction. I decided to celebrate with a trip to Tibet to learn meditation from the masters there, something I never would have been able to do when I was old, obese, and inflamed because it involved a lot of hiking and steep terrain.

I had just descended 7,500 vertical feet in one day in Nepal when I knew there was something terribly wrong with the cartilage in my knees. The cartilage itself was bruised from all that hiking, and I could barely walk across the street even using two trekking poles. I had exactly one week to recover before setting out on a rugged 26-mile walk at 18,000-feet elevation around Mount Kailash, which is considered to be the holiest mountain in the world. I knew that eating some extra collagen would be beneficial for my joints, but at the time collagen supplements didn't exist and there was no bone broth to be found in Tibet. I had to get creative.

The next day, the bus I was in stopped about halfway between Kathmandu and Lhasa in a town with only one restaurant. It had mud walls and a dirt floor and was filled with locals. I asked a Chinese friend from the bus to read the menu for me and quickly ascertained that the best source of collagen in the place was . . . pig's ears. Without hesitation, I ordered it, and a few minutes later I came face-to-face with a giant bowl of cold boiled pig's ears. I looked around to see if Joe Rogan, the host of *Fear Factor*, was hiding to challenge me to eat them for an absurd cash prize, but he was nowhere to be seen.

I had the idea that the pig's ears would somehow be more palatable if I could find a way to warm them up, so I ordered some watery soup and dipped the ears in one at a time before biting into their rubbery blandness. It was the second worst meal of my life. (The winner, during that same trip, was Chinese military ration sardines heated over a yak dung fire.) The pig's ears didn't have much taste, but the texture was wholly unappealing. However, I was shocked when I woke up the next morning and could walk without using trekking poles. Two days later, I could jog up a short hill. That is the magic of collagen. But I didn't want to have to eat pig's ears every time my knees hurt, so I worked hard to bring collagen to the market years later. I just couldn't see blending pig's ears into yak butter tea!

While I was in Tibet I met many old yet vital, energetic people and learned about their practices for pursuing a long, rich life. As I sat with meditation masters and Buddhist monks, I saw that a mind that can control its response to stress is the world's most advanced anti-aging technology. If you're walking around in a perfect environment eating all the right foods but your fight-or-flight response is always switched on like mine used to be, there is no doubt you'll age more quickly.

I made it to Mount Kailash thanks in part to the collagen in those pig's ears, but between the elevation and below-zero temperatures, I was hurting. Chilled, hypoxic, and exhausted, I staggered into a small guesthouse, where a kind Tibetan woman handed me a creamy cup of traditional yak butter tea. It was delicious, but more important, I felt like it brought me back to life. I even wrote about it in my travel journal. The air was still thin, but I was suddenly and remarkably full of energy, and I had to understand why. You're not supposed to want to dance when you're at 18,000 feet.

When I returned home I brewed some tea, tossed it in the blender with some butter, and was left with a greasy cup of tea that most certainly did not impart any mental clarity, unless you count the adrenaline from mild revulsion. Clearly, something different was happening back in Tibet. Figuring my problem was the tea, I spent a ridiculous $200 on a variety of high-end teas from a local Chinese merchant, but none of them had the magical effect I remembered. So I went to my local Whole Foods and another gourmet store, where I bought every single brand of butter from around the world to see if that was the variable that mattered. I tested twenty-four butters, and learned the trick was to use unsalted butter from grass-fed cows. You simply don't get the same results using butter from cows that eat corn and soy, because those oils end up in the butter, giving you more omega-6 fats. The yaks that provided the milk for the butter I had in Tibet certainly didn't eat any corn, because it doesn't grow there!

From my anti-aging work, I knew about the healthy fat in coconut oil, so I began experimenting with adding coconut milk and oil along with the butter, but the coconut flavor was too strong, and it didn't add any more energy than butter alone. So I switched from tea to coffee, my first love. The coffee stood up to the coconut oil better than tea, but the real magic happened when I switched from coconut oil to concentrated oil that is extracted from coconut oil called medium-chain triglyceride (MCT) oil. More than 50 percent of the fat in coconut oil comes from the different subtypes of medium-chain triglycerides. There are four types of MCT oils. All are flavorless, but the rare types convert effectively into ketones, your mitochondria's preferred fuel source. This was the genesis of Bulletproof Coffee.

The only problem was that MCT oil caused "disaster pants" even though it helped my brain. I should have bought stock in Charmin as

I worked through that problem . . . The solution was to remove certain types of MCT using triple distillation and then use a special filtration process, leaving only one type (eight-chain MCT), which became Brain Octane Oil. (Yes, I sell it. I use it. I give it to my kids. It works. Someone had to do it! It created a revolution in food.)

You may think that avoiding carbs or fasting for a few days are the only ways to enter ketosis (the state in which your body burns fat for fuel), but adding MCT or Brain Octane Oil to your diet hacks ketosis. Brain Octane turns into ketones when you consume it, even if carbs are present. Research that came out after I launched Brain Octane shows that it raises ketone levels four times more than coconut oil and twice as much as normal MCT oil.[23] In fact, the study says, "In healthy adults, C8 [the exact triple distilled version in Brain Octane] alone had the highest net ketogenic effect over 8 hours," and it could "help in developing ketogenic supplements designed to counteract deteriorating brain glucose uptake associated with aging."

Normal MCT oil is a conundrum for oil chemists. There are four different lengths of fats that are called MCT. All four are technically saturated fats, but unlike other saturated fats, your body won't use MCTs to make cell membranes. It's as if they are meant to be burned for energy. It is more accurate and useful to start calling MCTs "energy fats" instead of saturated fats. That's why I do not count MCT oil as a saturated fat and why you can laugh at anyone who says to avoid MCT because it's saturated. Sadly, the most abundant and cheap MCT, lauric acid, which makes up half of coconut oil, does not have these special energy powers.

To live longer and heal faster, I recommend adding either C8, its weaker cousin MCT, or its even weaker cousin coconut oil to your coffee, your salad dressings, smoothies, and so on. My kids love it drizzled on sushi! These "energy fats" do not count in the recommended ratios of fat in your anti-aging diet, as they will convert to energy instead of being stored on your body. These are extra/unlimited sources of fat. Also, when it comes to sourcing, I recommend purchasing MCT oil made from coconut oil, not palm oil. Most MCT is derived from palm oil, and palm deforestation poses a serious threat to the environment and kills orangutans. I switched to a coconut-derived MCT oil several years ago, because I simply couldn't imagine feeding oil to my kids that was created from practices that harm the environment they will inherit.

The discovery of using energy fats in the morning helped me benefit from autophagy because I was able to fast without getting cold or hangry (which, by the way, was added to the dictionary in 2018, the same year as *biohacking*). Because butter and MCT oil do not contain any appreciable quantity of protein, I was able to feel full and burn ketones while temporarily stressing my cells, which thought I was fasting and started recycling protein more rapidly. This boost in autophagy without hunger is one of the most profound benefits of Bulletproof Coffee. It is a permanent part of my quest to live to at least a hundred and eighty.

Yet, since I made my first cup in 2004, I've continued to discover more reasons why it works. To my surprise, one of them has to do with melanin, the pigment in your skin, which also exists in other parts of the body. When exposed to sunlight or mechanical vibration, new research indicates that melanin likely has the power to break apart water molecules, freeing up oxygen and electrons that your mitochondria can use to make energy.[24] Our bodies actually create melanin by linking together polyphenols, chemicals that occur naturally in plants. Polyphenols are packed with antioxidants and thus offer us a powerful defense against aging. The best ways to stimulate melanin production are to eat plenty of leafy green plants and herbs, drink tea and coffee, get adequate sun exposure, and exercise regularly.

This new information about melanin made me think back to my time in Tibet. I noticed that locals who carried all their belongings on the backs of yaks made sure to always have blenders hooked to portable batteries just to make yak butter tea. They were clearly onto something. Tea and coffee contain large amounts of polyphenols. Coffee also contains melanin and similar compounds called melanoids. Is it possible that Bulletproof Coffee and yak butter tea are so energizing because the mechanical vibrations from the blender break up the melanin and melanoids,[25] providing free oxygen and electrons for your mitochondria? Is this why the yak butter tea made me feel so much better in high altitudes where there was less oxygen? I think so.

COFFEE + TIME = KETONES

Recently I interviewed Satchin Panda, a leading researcher on circadian rhythms, the natural twenty-four-hour cycles of all living beings, and learned something new about Bulletproof Coffee. According to Satchin, it's part of our natural rhythm to start producing ketones at the end of our fasting cycle. For most of us, that would be in the morning before we *break our fast* with the aptly named meal, breakfast.

Those ketones have a huge impact on our cardiovascular and brain health. Satchin observed that when mice produce ketones toward the end of their fasting cycle, those ketones go directly to brain cells called *clock neurons*, which monitor the environment in the brain and help to regulate circadian rhythm. When ketones reach those clock neurons, they receive a signal to become awake and alert and begin what is called exploratory activity. Of course exploratory activity is more pleasant than desperately wanting to hit the snooze button in the morning.

This makes perfect sense from an evolutionary perspective. Just a couple hundred years ago, our ancestors fasted all night and then had to hunt for food in the morning. Their brains and muscles had to work really well in that hungry state in order to successfully find food, and ketones were the answer. This is why we are programmed to build up ketones during the last couple of hours of our fasting period. Those ketones give our brains, muscles, and hearts more energy so we can hunt—exactly what Satchin has seen in his lab rats. An hour or two before they were fed in the morning, they got up and started looking around, exploring, and getting ready to hunt.

The problem is that most people don't fast long enough to take full advantage of this biological phenomenon. According to Satchin, there are tremendous health benefits to extending our daily (or nightly) fast. He says that when people limit their eating window to ten hours and make no other dietary changes, they see reductions in inflammation levels, triglyceride levels, and cancer risk, along with improvements in sleep within weeks. Is this because of the natural boost in ketones or because intermittent fasting boosts autophagy—or both?

But remember, you do better when you practice ketosis intermittently. Staying in ketosis for long periods of time compromises your metabolic flexibility—your body's ability to burn either glucose or ketones for fuel. Maintaining metabolic flexibility is incredibly important for your longevity. There are two states your body must be able to handle effortlessly. The first is periods with ketones and no carbs, and the second is periods with carbs and no ketones. To gain metabolic flexibility, the best thing you can do is cycle in and out of ketosis every week. To do this, you limit carbohydrate intake most days, and on one to two days per week you eat low-sugar carbs. While this works for die-hard biohackers, most people enjoy eating more carbs. With the power of technology, it is possible to have both ketones and carbohydrates present in your body at the same time, which can also generate metabolic flexibility. To do this, eat moderate low-sugar carbohydrates like white rice or sweet potatoes, and at the same time consume lots of energy fats. That way, you'll have some ketones present for your neurons and some glucose present for your brain's maintenance cells. Most people find this more sustainable than a pure cyclical ketogenic diet, but both work.

There is no doubt that strategies like ketosis, intermittent fasting, and the maintenance of a healthy circadian rhythm play a critical role in our longevity. This leads to the next essential step on our quest to become Super Human—and that is getting enough highly efficient, good quality sleep.

Bottom Line

Want to not die? Do these things right now:

• Avoid all conventionally grown grains, produce, and animal products. Even better, skip grains altogether and opt for tons of organic vegetables, limited organic fruit, and meat from pastured animals.

• Don't eat fried stuff. Ever.

• Eat enough protein (from pastured animals, eggs, wild fish, or nonallergenic plants) for tissue repair and an additional 20 plus grams of grass-fed collagen, and don't fry, char, blacken, or barbecue meat (sorry). For lean people, that's 0.5 grams per pound of body weight. For obese people, that's about 0.35 grams per pound of body weight. For pregnant women, elderly folks, or athletes, it's 0.6 grams per pound.

• No matter how much fat or how little fat you eat, eat the right ratios. Lean people eat about 50 percent saturated, 25 percent monounsaturated, and 15 to 20 percent *undamaged* omega-6 and 5 to 10 percent omega-3, including EPA and DHA. If you are fat like I used to be and want to live like a Super Human, eat 50 to 70 percent saturated, 25 to 30 percent monounsaturated, and only 10 percent *undamaged* omega-3 and omega-6, with added EPA and DHA so that you eat more omega-3 than omega-6.

• On some days, limit your eating window to eight to ten hours a day based on what works best for your schedule. Good options are 12:00 P.M.–8:00 P.M., 9:00 A.M.–5:00 P.M., or 10:00 A.M.–7:00 P.M. Have breakfast sometimes, especially if you're tired or stressed. Don't eat after dark.

• Teach your metabolism to be flexible by having ketones present in your system every week. Practice a cyclical ketogenic diet by fasting, avoiding carbohydrates for a few days, or adding "energy fats" to your food (or coffee) that convert directly to ketones.

SLEEP OR DIE

Sleeping feels good, but ever since I was a kid, there was always something more fascinating and productive I'd rather do than go to bed. I resented having to dedicate so many hours each day to something I saw as basically a waste of time. So for most of my life I skimped on sleep. Even the first two years after founding Bulletproof, I slept for about four hours a night, at most a self-imposed five hours. I used the extra three hours a day to be a father, start Bulletproof, and still pay the bills with my day job.

My sleep deficit almost certainly contributed to the diseases of aging I faced as a young man. It turns out that lack of quality sleep doesn't just leave you tired and unable to perform in the moment; it also rapidly accelerates aging. The good news is that you can learn to be a Super Human sleeper and cram more high quality sleep into fewer hours and still get all the benefits. For the past five years, I have been getting progressively healthier, leaner, and younger on six hours and five minutes of sleep a night, but I use every technique in this chapter to sleep like a professional.

Perhaps you will choose to get more sleep than I do. Regardless of how many hours you sleep, the information in this chapter is intended to help you make the most of the sleep you do get. It doesn't matter how old you are, how busy you are, or how much money you have. Sleep is the ultimate tool to sharpen every skill and add more quality years to your life. So get better at it.

HOW LACK OF SLEEP WILL KILL YOU

Like it or not, a lack of *good* sleep directly increases your risk of dying from one of the Four Killers. Meanwhile, just one good night of sleep can improve your ability to learn new motor skills by 20 percent,[1] and getting regular quality sleep increases your ability to gain new insight into complex problems by 50 percent.[2] This improved brain function could potentially help ward off cognitive decline with age and is befitting of a true Super Human. Good quality sleep also promotes skin health and youthful appearance,[3] controls optimal insulin secretion[4] (making you less likely to develop diabetes), and encourages healthy cell division.[5] Sleep is an essential strategy in protecting against all Seven Pillars of Aging.

In the previous chapter we discussed Satchin Panda's research on longevity and circadian rhythms. As part of my research for this book I went to his lab and had a great time with his PhD students looking at how the combination of food, light, and too little sleep affected rats. They walked me through new research showing that eating late at night dramatically reduced the quality of the rats' sleep and that poor sleep impacted the rats' ability to control blood sugar by up to 50 percent. That's huge! In fact, it's more than what most medications can do.

In rats and humans, the pancreas is responsible for making insulin. Satchin has studied insulin-producing cells in the pancreas and found that they, too, have their own circadian rhythm. At night when melatonin, a hormone that helps regulate wake and sleep cycles, is released, insulin-producing cells shut down, too. So if you eat something sugary late at night, your body's insulin response is not as effective as usual. So that late-night piece of cake leads to a blood sugar spike and then a crash that triggers the release of adrenaline . . . which keeps you awake at three A.M.

If you get less than six hours of sleep, the hormones that control how hungry and/or satiated you feel (ghrelin and leptin, respectively) start to work against you. Ghrelin increases, making you feel hungry, and leptin decreases, making it more difficult for you to feel satiated. This is one reason that sleep loss leads to obesity and all the many health problems that go along with it.[6]

Sleep is also incredibly important for warding off Alzheimer's disease, the killer many of us fear most as aging begins its silent creep. When you are asleep, your brain undergoes a natural detoxification process. The glymphatic system, a waste clearance pathway comparable to the lymphatic system, which drains fluids from tissues in the body, sends cerebral spinal fluid through the brain's tissue and flushes out cellular waste and neurotoxins.[7]

This is a big deal, as the glymphatic system clears out the amyloid proteins that are the hallmark of Alzheimer's when they build up in the brain. While we don't yet have hard evidence that Alzheimer's disease is caused by a lack of sleep and thus not enough time for the glymphatic system to work its magic, I would wager that it's a contributing factor. In fact there is some evidence of this. A small study on twenty human participants showed that losing just one night of sleep causes an increase in amyloid proteins in the brain.[8] That may be a small sample, but it's enough to convince me to make sure my glymphatic system has a chance to fully detox my brain each night. That doesn't mean sleeping for eight hours; it means sleeping like a boss.

Since mitochondria play a role in the glymphatic system process and sleep in general, everything you do to strengthen your mitochondria can also help you sleep better and thus keep your brain clear of amyloid plaques. There are also simple things you can do to enhance your glymphatic system function. For example, studies on rats show that sleeping on one's side improves glymphatic clearance compared to sleeping on the stomach or the back.[9] While we don't have studies proving that this transfers to humans, we know that side-sleeping humans have lower blood pressure and heart rate.[10] Sadly, they also get more vertical wrinkles than back sleepers, but sleeping on your back increases your risk of sleep apnea, a condition in which the upper airway becomes blocked during sleep. Sleeping on your back will make you less wrinkled but more likely to die. Not a great trade-off. I'd opt to stay alive and hit those wrinkles with other hacks in this book.

Apnea in and of itself puts you at a much higher risk of dying from one of the Four Killers. Sleep apnea is often the result of dysfunctional mitochondria, and it can be deadly.[11] If you snore, your risk of developing diabetes, obesity, and high blood pressure is nearly double that of someone who does not. And if you snore *and* you wake up feeling groggy and/or have trouble falling asleep, your risk goes up 70 to 80 percent, respectively.[12]

As you read earlier, bad quality sleep causes poor blood sugar regulation. It's also true that dysfunctional mitochondria cause bad sleep, which then causes poor blood sugar regulation! No matter how you slice it, if you don't get enough good quality sleep, you will age faster and die sooner. Which begs the question . . .

HOW MUCH IS ENOUGH?

When I learned about how critical good quality sleep is to aging well, my perspective on sleep changed for good. Instead of seeing it as something to skimp on, I made it my goal to hack my sleep so I could get all of the benefits of a good night's sleep without having to sacrifice eight hours of my life every night. Some of these efforts have been more successful than others.

In the year 2000, when Google was just eighteen months old, an early biohacker posted the Uberman Sleep Schedule in a dark corner of the Internet. This was the first writing to propose that you could get away with only three hours of sleep per day as long as you were willing to sleep in several carefully timed, precise naps at exactly the same times each day. This technique is now called polyphasic sleep.

Intrigued by the approximately eleven years of my life I'd reclaim from this schedule, I tried it. The amount of time and energy it takes to do this is absurd, not to mention the social and professional interruption from napping at the same time every day and feeling wrecked if you miss one nap. Polyphasic sleep is not compatible with having either a career or a social life. Some people have success with it, but personally I felt like an unproductive, antisocial zombie. The idea of getting by on a couple of hours of sleep at a time is a beautiful dream (get it?), but it just didn't work. I was starting to feel resigned to having to sleep eight hours a night . . .

Then I came across a study from the Keck School of Medicine of

USC and the American Cancer Society that looked at over 1 million adults ranging in age from thirty to a hundred and two and correlated how much they slept with their mortality rates.[13] The results of this study changed the way I thought about sleep forever. The data was actually collected in the 1980s, but it was so complex, showing differences in outcomes with just a half-hour difference in sleep length, that they couldn't crunch it all with 1980s computing, so the information sat there for years until researchers could use high-speed computing. The researchers found that the people who lived the longest slept six and a half hours a night, while people who slept eight hours a night consistently died more from any cause. Ha! Take *that*, all you doctors who told me I had to sleep at least eight hours each night!

You might hear this and draw the conclusion that in order to live longer you should simply sleep less, but that is unfortunately the wrong conclusion. What you can take away from that study instead is the fact that the people who lived the longest were the healthiest people. They required less sleep because they didn't need as much time to recover from chronic illness, inflammation, and/or everyday stress. If aging is "death by a thousand cuts," sleep equals recovery from many of those "cuts." The fewer cuts you need to recover from, the less sleep you need.

I started using my sleep length and corresponding energy levels to measure whether I was doing things during the day that made me older. I knew that if I jumped out of bed ready to bring it after six hours of sleep, I was on the right track. But if I felt groggy after a solid eight hours of sleep, that meant I was probably doing something that made me sick and inflamed. This explains why I needed less sleep when I started following the Bulletproof Diet. I was taking fewer hits from the foods I ate, so I didn't need as much recovery time.

This became a two-step process. Step one: Reduce the number of hits I took so my body required less recovery time. Step two: Increase the return on my sleep investment by improving its quality. Bottom line—if you're healthy enough, you can use sleep strategically as a performance-enhancing drug instead of a drag. You still have to get enough sleep, but the other hacks you'll use to become Super Human will reduce the number of hours of rest you actually need.

HOW WELL DID YOU RECOVER LAST NIGHT?

In order to work on improving the quality of my sleep, I began a long journey of understanding my sleep, a journey that is still going strong after nineteen years. There are all sorts of reasons to pay attention to your sleep. If sleep is recovery, you need to know how well you recovered last night so you can make an informed choice about what actions to take today. For instance, if you know you slept poorly, a heavy workout will age you instead of making you stronger; a high-sugar meal will impact your blood sugar even more than usual; and even small amounts of stress will be damaging.

Quality sleep is like having money in your recovery bank account. Can you imagine not checking your bank account on a regular basis? If you can see where your sleep stands today, you can zero in on small changes you can make to improve your sleep, recover better, and stay young tomorrow.

In 2004, I was finishing a brutal two years that had me working full time while enrolled at an Ivy League business school. Sleep was in short supply, as you'd imagine. So I became one of the first purchasers of an expensive headband that tracked my sleep and told me exactly how well I did every night. The data was enlightening and helped inform many of my early biohacking practices. Unfortunately, Victoria's Secret definitely did not approve of these early tracking devices (and neither did my wife). Thankfully, there has been quite an evolution in the quality and attractiveness of sleep trackers since then.

Seven years later, I became the chief technology officer for a wristband sleep and exercise tracking company called Basis (which Intel has since acquired). Before most people were wearing wrist trackers, I was able to track my sleep, make strategic changes, and get more out of the time I spent with my eyes closed. In fact, I've purchased and tried just about every sleep tracker on the market. A sleep tracker is an anti-aging device with one of the highest ROIs. I promise that you have no idea what your brain is doing while you sleep. Before we get to what technology to use, ranging from free to a few hundred dollars, it's important to know what you are looking for when you track your sleep.

SLEEP BASICS

Of course you want to know exactly what time you fell asleep, what time you woke up, and how this information varies over time. Did it take you a long time to fall asleep after you went to bed? Did you wake up several times during the night even if you don't remember them? Are you wasting your night with light sleep? These are all important factors in determining the quality of your sleep. When I did the crazy "zero carbs for nineteen days" experiment, my unattractive headband sleep tracker showed me I was waking up eight to twelve times every night, yet I had no recollection of waking at all. I did feel like a zombie in the morning, though. It was my sleep data that eventually made me quit that experiment!

It's also worth paying attention to snoring when tracking your sleep for all the reasons mentioned above, particularly because snoring is a sign of inflammation. I used to snore terribly because the back of my throat was inflamed and partially blocked my airway. Now I don't normally snore more than a couple of minutes a night, and I am usually able to connect it to something I ate the day before that caused inflammation. I also get a handy recording of my snoring so I can't deny that it's happening! This is incredibly valuable information because food that inflames your throat also causes aging inflammation throughout your body.

REM/SLOW-WAVE SLEEP

When you're sleeping, you cycle through two types of sleep each night—rapid eye movement (REM) sleep, which is when you dream, and non-REM (NREM) sleep. NREM sleep comes in three flavors: crappy (stage 1 useless light sleep), decent (stage 2 middle sleep that is still considered light sleep), and awesome (stage 3 deep delta sleep). To age or perform like a Super Human, it's your job to spend as much time as possible in deep or slow-wave delta sleep. This is when your breathing and heart rate drop to their lowest levels and your brain waves slow down and get wider (as measured by a test called an electroencephalogram, or EEG). These slow waves are known as delta waves, and your brain produces them at a frequency of 1 to 4 hertz, a unit of measurement that is equal to one per second. To put that into context, gamma waves, the fastest brain waves, have an average frequency of above 40 hertz.

It's important to spend a lot of time in slow-wave sleep because it helps the brain recover from all it learned during the day.[14] It strengthens memory and helps with memory consolidation, when the brain turns short-term memories into long-term ones.[15] This can help young people perform better at school and work and help old people avoid the memory loss that too often comes with age. Slow-wave sleep also reduces levels of the stress hormone cortisol and triggers the release of hormones like prolactin and growth hormone, which together support the immune system.[16]

If you want to stay young, get more deep sleep. There is unfortunately a big decrease in deep sleep between your teenage years and your mid-twenties, and your body substitutes middle sleep for the deep sleep you lose. As you get older, slow-wave sleep continues to decline . . . unless you do something about it. Only nineteen years ago, researchers examined six hundred sleep studies and reported that total sleep time and percentage of deep sleep decline with age. For every decade you live past thirty, subtract 12.2 minutes from your total sleep (at least if you're average, not Super Human). Even worse, your amount of useless light sleep increases to over 50 percent. That sucks if you want to live to at least a hundred and eighty, and it sucks if you want those three to four hours of wasted sleep available to do fun stuff. Your amount of REM sleep will mostly depend on your health status up until the age of sixty, when you'll start getting less REM sleep unless you do something to stop it.[17]

When it comes to deep sleep, teens need 1.7 to 2 hours, and people over the age of eighteen need 1.5 to 1.8 hours.[18] The odds of your actually getting that much sleep are not very good, but until you measure it, you won't know how much sleep you're getting. Schools force teenagers to wake up incredibly early, despite the fact that they need to sleep until at least eight A.M. to get enough deep sleep. This is a critical measure of health—one that matters much more than how fast you run, what your VO_2 max is, how much weight you can bench-press, or how your abs look. At age forty-six, I sleep like a teenager. I can achieve two or more hours of deep sleep and two to three hours of

Sleep score from my Oura Ring sleep tracker showing more deep sleep and REM sleep than teenagers get in eight to ten hours, even though I slept less than six hours. I also used the Sonic Sleep Coach app, TrueDark sleep glasses, and supplements to reach these levels.

REM sleep in six to seven hours if I use the sleep hacks we'll discuss later in this chapter.

The more time you spend in either REM or delta sleep, the more restorative your sleep will be. That means you can spend less time asleep and wake up more refreshed, younger, and smarter. You know, Super Human.

HEART RATE VARIABILITY (HRV)

Your autonomic nervous system regulates bodily functions such as metabolism, breathing, and sleep. It is made up of two systems: your sympathetic nervous system and your parasympathetic nervous system. The former is responsible for your stress response—it initiates your highly stimulating fight-or-flight response in the face of a stressor. The parasympathetic branch activates activities associated with rest and recovery such as sexual function and digestion.

When your sympathetic nervous system is activated, your heart rate becomes very even no matter how fast it beats. This is a sign of a stressed animal. But when your parasympathetic nervous system is in charge, there is a greater variability between heartbeats. In other words, you may have the same number of beats per minute, but they are less rhythmic when you are relaxed. You are able to recover from stress more quickly, and this is reflected in a quick acceleration and then deceleration of the heart rate. Your heart rate variability (HRV) is the measure of this variability between heartbeats. Low HRV is associated with anxiety disorders and even cardiovascular disease, while high HRV is linked to cardiac resilience and overall heart health.

There is also a direct relationship between HRV and sleep. When you do not get adequate quality sleep, your body becomes stressed and the sympathetic nervous system is activated. This reduces HRV. One study from the University of Pennsylvania found that just five nights of sleep restriction significantly reduced participants' HRV.[19] In contrast, consciously increasing your HRV during the day improves your sleep efficiency.[20]

Here's the bad news. Lots of things, including the aging process, lower your HRV. Being out of shape, overtraining, chronic stress at work or home, inflammation, and infections all reduce HRV. The good news is that even if you don't measure your HRV, you can usually increase it by meditating, doing breathing exercises, taking a hot bath, sleeping better, eating better, consuming fewer toxins (including alcohol), and even taking the right supplements for your biology. I often find type A personalities—executives who also want to be endurance athletes—have the HRV of very old people. When they actually see the results, they realize that burning the candle at both ends isn't serving them.

When you decide to track your HRV, you are effectively measuring how stressed you are on a physiological level. This is incredibly valuable information. You already know that stress from all sources accelerates aging. So when you know your body is stressed, you must choose to actively recover rather than burdening your system even more. This alone can dramatically alter the course of your aging.

TRACKING DEVICES

We have come a long way since the early days of 1970s clip-on pedometers, home bathroom scales, and even the old Basis tracker I worked on. The new sleep tracking devices are sleeker, more attractive, and more accurate, and measure everything from your HRV to specific sleep states and even brain waves. Some can help you wake up more gently so you are not jarred out of sleep and forced to start your day in a state of physiological stress.

Unfortunately, most of the wristband devices that claim to track your sleep don't gather particularly great sleep data because they were designed to track movements, not sleep. When the industry realized there is little value in movement data, they tried to add sleep functionality. It's not really their fault; it's been this way since the 1970s. We've all heard the "ten thousand steps per day" metric, but did you know there is no science behind this? In 1965, a Japanese company, Yamasa Tokei Keiki, created a clip-on pedometer designed to get people to walk ten thousand steps per day and simply made this number up.

This is why your best options to track your sleep are either a simple, inexpensive app on your phone or an unobtrusive high-tech ring. There are plenty of options in between, with a wide range of data. Your mileage may vary. If you already have a device, you can likely use its limited sleep tracking to see if you can improve your sleep, but most devices don't provide a great window into how much deep sleep you're really getting, and even fewer show precious HRV data.

No matter which sleep tracker you choose, the important thing is to gain a sense of how well you are sleeping and then begin paying attention to the factors that influence the quality of your sleep.

You can make smarter choices about how to spend your energy the next day when you know how much savings you have in your recovery bank. To start that process, I recommend the following technologies.

SLEEP CYCLE APP

The Sleep Cycle app has the highest return on investment for sleep tracking, but not because it gives you great data. It doesn't. The ROI is high simply because it doesn't cost anything except the time you spend to set an alarm clock you already use! It is a simple, inexpensive option to roughly track your approximate sleep cycles using only your phone's microphone and a movement sensor. It analyzes breathing and movements to determine when you're awake, in deep sleep, or in REM sleep. It will (embarrassingly) record your snoring and graph how many minutes you snore per night. Best of all, its alarm feature makes sure not to interrupt your deep sleep. You tell the app when you want to wake up, and then you place your phone (always in airplane mode) next to your bed or on your nightstand when you hit the sack. When you're in a light sleep phase around your programmed wake-up time, your alarm will go off. This way, you still get to complete your cycle of deep sleep and you're not as tempted to hit the snooze button. Sadly, there is no measure of heart rate variability, and the sleep data is not highly accurate.

I have used this app for more than five years almost every single night because the alarm never jolts me from delta sleep, and I like the snoring reporting!

COST: Free, or $29.99 per year for Premium

PLATFORMS: iOS, Android

MEASURES:

- Intelligent wake-up
- Sleep analysis
- Nightly sleep graph
- Alarm melodies
- Snooze

- Apple Health integration (iPhone only)
- Database export (iPhone only)

OURA RING

The Oura Ring is my number one sleep tracking device recommendation. While it isn't the cheapest, it's not the most expensive, either. It does provide the very best data of any sleep technology and is easy to use. With a full range of metrics and advanced analyses, the Oura helps you see what's happening under the hood while you're sleeping. The more data you can track, the more insight you have, and the easier it is to make small changes that will help you get the best rest in the least time. The Oura Ring is also one of the most attractive wearable devices out there. It's sleek and modern and looks like a regular ring, not a clunky tracking device.

The Oura offers a unique feedback feature that analyzes your data to calculate a Readiness Score—an assessment meant to help you understand whether you're ready for a hard-charging day ahead or if you should focus on rest and recovery. The Readiness Score uses information from your sleep the night before, your activity the day before, and a handful of other measures to help you make informed decisions about how you should plan your day.

This ring also tracks your average body temperature throughout the night, which can help you understand if tweaking the temperature in your bedroom will improve your sleep. My data shows that, indeed, a cooler room equals more REM sleep. Information about body temperature is also useful for women who are tracking their menstrual cycles. You can expect to see a bump of about 0.3 degrees right around the middle of your menstrual cycle when your progesterone levels rise during ovulation. This is valuable information for women who want to take control of their biology.

The Oura is one of the few trackers that provide your HRV and respiratory rate. These are good indicators of how stressed you are and therefore how quickly you are aging. As you begin to follow the recommendations in this book, you can check in on these measures to see how they change. I genuinely enjoy looking at my Oura score in the morning to see how well I did (or didn't do) the night before. It is my most trusted, useful sleep tracker and overall biomonitor, and it tracks exercise information, too.

COST: $299 (for the basic model)
PLATFORMS: iOS, Android
MEASURES:

- Total sleep
- Sleep efficiency
- REM sleep
- Deep sleep
- Light sleep
- Latency (time from pillow to falling asleep)
- Timing
- Body temperature
- Heart rate variability
- Respiratory rate
- Calorie burn
- Steps

(Disclosure: After completing the first draft of this book, I joined Oura's advisory board and became a small investor in my favorite sleep app because it's the best I've used.)

After tracking my sleep for about two decades, I have learned to correlate how I feel with the quality of my sleep. When I wake up, I always guess what my Oura Ring score is based on how I'm feeling. And I've found that over time those guesses have gotten closer and closer to the Oura Ring results because I've learned to detect precisely how it feels to get a night of good quality sleep—and how it feels when I don't. I can also use this information to test new sleep hacks and see which gives the best return on investment. When I make a tweak to my sleep schedule, I track how those differences stack up.

This has allowed me to fine-tune my protocols to the point where I can travel multiple times a week and no longer have to deal with jet lag. This is an art and a science because everyone's biology reacts to different inputs in unique ways. For instance, you might not be as sensitive to blue light as the next person, or you might be extremely sensitive to blue light. The point is that you don't always know what's killing you until it's too late. Mastering your sleep is an important way to eliminate some of the cuts that are aging you and take control of your biology.

SLEEP BETTER, NOT MORE

Tracking your sleep is only half of the equation. You can have the best, most accurate data in the word, but if you don't do anything with it, it's worthless. Taking action is equally important. The following hacks range from the extremely simple to somewhat complex, and from free to moderately expensive. Choose those that are best for you to make sure you are getting the greatest possible return on your sleep investment.

DIRECTLY DIAL IN YOUR SLEEP STATE

Since delta (slow) waves are so good for the brain, scientists have found ways to create more of them during sleep. In one study, transcranial magnetic stimulation (TMS)—which uses magnetic fields to stimulate parts of the brain—triggered the release of slow waves that spread to the rest of the brain,[21] but that's not an option for most people on a nightly basis. Fortunately, there is an easier way to increase delta waves. In a 2010 study, researchers experimented with sound and found that short auditory tones—50 milliseconds each played at a rate that mimicked the natural changes in brain neurons during sleep—increased slow waves by nearly 50 percent during non-REM sleep.[22] Other studies found similar results.[23]

Fortunately, there's an app for that! It's called Sonic Sleep Coach, created by Daniel Gartenberg, PhD, who has been awarded more than $1 million in NIH grants to study sleep. Sonic Sleep Coach uses your smartphone's microphone to sense your breathing while you sleep, which indicates when you're in deep or REM sleep. It plays specific sounds to increase REM when you're already dreaming and different sounds to increase deep sleep. The result is a meaningful improvement in your sleep efficiency. If it picks up on noise, it even plays audio tones to block out city sounds so your sleep isn't disturbed. I always use Sonic Sleep Coach when I'm on the road, and sometimes at home.

(Disclosure: After writing this, I became an advisor and a very small investor in this impressive app because I believe in improving sleep, and it works!)

REDUCE BLUE LIGHT

Other than a cup of coffee right before bed, nothing is more like kryptonite for your sleep than bright blue or white light in the evening. It truly makes you old in several different ways. Blue light is everywhere—we get normal amounts from the sun, but we get large, unbalanced doses from light-emitting diodes (LEDs) used in energy-efficient bulbs and to illuminate the screens on TVs, computers, tablets, and smartphones. Blue light has a short wavelength, so it produces more energy than light frequencies with a longer wavelength, like red light. The odds are high you've heard some of this already but dismissed how big of a problem it is on your path to becoming Super Human. The data is convincing, and it's easier than you think to transform blue light into only a minor problem.

Blue light is not all bad. Exposure to blue light during the day wakes you up, makes you more alert, and can even improve your mood. White- or blue-light emitting goggles and panels are used to treat a number of issues such as seasonal affective disorder (SAD), jet lag, and premenstrual syndrome.[24] The problem is that newer artificial lights like LEDs and compact fluorescent light (CFL) bulbs don't contain most of the infrared, violet, and red light that's found in sunlight. Instead, they increase the intensity of blue light to a level that our eyes, brains, and bodies haven't evolved to handle. This is what I call "junk light" because it's just as unhealthy and aging as junk food.

You're bombarded with junk light throughout the day and for much of the night—when you're on your phone, working at your computer, or watching TV—and all this blue light exposure messes with your sleep.[25] Blue light shifts your circadian rhythm in part by suppressing melatonin, the hormone that tells your brain when it's time to sleep. This tricks your body into thinking it's daytime 24/7.

Normally, the pineal gland, a pea-sized gland in the brain, starts releasing melatonin a couple of hours before it's time to go to bed. But blue light can mess with this process by stimulating a type of light sensor called intrinsically photosensitive retinal ganglion cells (ipRGCs) in the retina of the eye. These sensors send light information to the circadian clock, telling the body when it's time to sleep and wake up using more than just melatonin.[26]

When light sensors are stimulated by blue light at night, you have a harder time falling asleep. A 2014 study found that people who read from a light-emitting device before bed took longer to fall asleep, slept less deeply, and were more alert than people who read a printed book.[27]

The amount of blue light you're exposed to at night has also been linked to rapid aging. The mitochondria in your eyes have to produce a lot more energy than normal to process blue light. When the mitochondria in your eyes are overtaxed, the rest of your mitochondria get stressed, too. This causes metabolic problems and inflammation throughout the body, which of course increases your risk of the Four Killers.

One study found that adults who were exposed to blue light while eating in the evening had higher glucose levels, slower metabolisms, and more insulin resistance compared to adults who ate in dim light.[28] Get some candles or at least a dimmer switch, like I have in my dining room. It's way cheaper than diabetes.

People exposed to high levels of outdoor blue light at night also have a higher risk of developing breast cancer and prostate cancer, compared to those who had less exposure.[29] Other studies have found that a disrupted circadian clock increases your risk of cancer.[30] Blue light exposure is also linked to obesity and metabolic disorders, which are both significant risk factors for cardiovascular disease.

Blue light can also lead to macular degeneration—damage to the retina that often leads to vision loss.[31] More than 11 million people over the age of sixty, including my father, have some form of macular degeneration, so it's an issue that hits close to home.[32] If you'd like to join me on my quest to live beyond a hundred and eighty, you'll want to be able to see as you age. Cutting out some blue light is not optional, unless you're okay with becoming one of the 100 million plus cases of macular degeneration I'm forecasting over the next decades.

We know that macular degeneration is a mitochondrial disorder and possibly also a blood coagulation disorder. To keep your eyes working, use all the mitochondrial hacks in this book, and make sure your blood is as thin as it should be. Fish oil, or better still, krill and fish roe oil, and turmeric are some anti-inflammatory supplements that can help. I take those and the eye-specific formula I created for Bulletproof and do every one of the items on the following list at least some of the time.

After a recent very detailed eye exam conducted by a leader in the field, I received some fantastic news: "Dave, even though you're forty-six, your eyes are exhibiting none of the signs of reduced flexibility expected for your age. You're 20/15 in both eyes, and can read the very finest print on the test. It's probably because of your glasses [True-Dark half–blue blocking], your diet, and your supplements." *Boom.*

Even if blindness doesn't worry you, you must reduce your blue light exposure in order to reduce your risk of dying. Here's how:

- **Unplug or cover unnecessary electrical devices in your bedroom.** Go through your room and unplug or tape over LEDs to black out your sleep area. I carry special die-cut dots (or sometimes electrical tape) with me when I stay in hotels so I can tape over the omnipresent blue LEDs on televisions, air conditioners, and alarm clocks.

- **Invest in blackout curtains.** If you try just one sleep hack, make it this one—it's a serious game changer. A dark room equals better sleep. Even with blackout curtains, watch out for the light that seeps in all around the edges of the curtains. Buy some Velcro and tape down the sides and put a valance on top, or just use foil over your windows. No, it's not exactly stylish, but it's a fair trade-off for quality sleep in my book. If you're working with a window-covering company, use these words: "This job will be a failure if I can see any light at all coming in past the blackout shades."

- **Switch to amber or red bulbs at night, or at least get a dimmer switch for your existing lights, and use it.** I keep a lamp with a red LED bulb by the bed.

- **Toss out bright white LED and compact fluorescent lights.** They have up to five times more blue light than incandescent or halogen lights even though they are cheaper to operate. You'll have less eyestrain, which you read above can help with your metabolism, and you'll have less chance of macular degeneration later.

- **Use glasses to protect your sleep.** I started wearing silly-looking blue blocker glasses in 2008. The first time I wore them onstage at a tech conference, I felt like the biggest dork ever, but my brain was so much happier that it was worth it. The funny thing is, I got about ten times as many business cards as usual at the conference because everyone recognized me as the guy in the yellow glasses. Based on what we now know about blue light, it's a bad idea to wear blue blockers during the morning and midday, because with *no* blue light, your body doesn't know it's daytime. That's why I started TrueDark, which makes daytime glasses that block only some blue light. (Yes, I believe in this so much I started a company to solve the problem. You don't have to wear TrueDark glasses to learn from this book!) For the best results, wear the amber ones you often see me wearing. If you want to hide your Super Human proclivities, there are now "invisible" TrueDark glasses you can wear to block some blue and maintain your secret identity. There are also patented glasses that block even more blue light that I wear a few minutes before bed—I do not achieve my best deep sleep scores without them.

- **Switch to Night Shift mode on your iPhone, iPad, and iPod Touch and leave it on all day.** While it's best you put all your devices down before bed, that's not always a good trade-off in this technology-driven world. Apple has a cool, relatively unknown hack to adjust the colors of your screen to warmer tones. It's called Night Shift, and it takes just a minute to set up. Swipe up from the bottom edge of any screen and press firmly on the brightness icon. Then tap the Night Shift icon to turn it on or off. Once you've done that, go to Settings > Display & Brightness > Night Shift. Here you can schedule when you want Night Shift to turn on and off (from sunset to sunrise, for example). You can also adjust the color to be warmer.

- **Install light filter apps.** On any computer or on any Android phone, you can use apps like f.lux or Iris, which adjust your display's color temperature depending on the time of day.

- **Shut down all screens two hours before bed.** This isn't always realistic, but when you can, do it. (I don't do this one often—the book you're reading was written at night.)

- **Take carotenoid supplements.** Your eyes need macular carotenoids—pigments that act as antioxidants—to protect them from junk light. The specific carotenoids lutein, zeaxanthin, and astaxanthin work together to protect the retina and reduce oxidative stress caused by blue light.[33] They are often sold together in supplements meant to support eye health, and of course I included them in Bulletproof's Eye Armor supplement.

- **Increase your exposure to high quality light sources before noon.** Make sure you spend some time outdoors in the sun to balance your exposure to artificial junk light. Fifteen to twenty minutes of natural sunlight each day is ideal, but you can also use a bright halogen 300-watt light. This is a simple way to boost your energy and mental performance now and halt aging in its tracks for long-term benefit.

It turns out that junk light does more than just destroy your sleep. It also directly messes with your mitochondria and creates inflammation, aging you every day. At the same time, there are beneficial light therapies that can prevent and even reverse some of this damage. So how quickly you travel down the path of aging is greatly determined by how you choose to light it. This leads us to the next chapter . . .

Bottom Line

Want to not die? Do these things right now:

• Download the sleep app of your choice (or buy a device) to begin tracking your sleep so you can see how well you are recovering each night.

- Best app: Sonic Sleep Coach.
- Best device: Oura Ring.

• When you know you didn't sleep well the night before, choose an activity that helps you recover instead of something that will stress your body even more. For instance, choose yoga over a heavy lifting day at the gym.

• Improve your sleep hygiene. Keep the temperature cool (around 68 degrees), use blackout curtains, and develop a nighttime routine so your body knows when it's time to wind down for the night.

• Reduce your blue light exposure at night any way you can. Dimmer switches, red LED bulbs, and special glasses are the best choices.

USING LIGHT TO GAIN SUPERPOWERS

Quick quiz: What gave Superman his superpowers? His cape? His Fortress of Solitude? Nope, it was actually the Earth's sunlight. With the right knowledge, you, too, can gain superpowers from the sun. And while you still won't be able to leap tall buildings in a single bound, you will have more energy, look better, and live longer.

Light is fundamental to our existence, yet most of us hardly even notice it. We are accustomed to using light as a simple tool to see more clearly after dark. But it is actually much more complicated than that. Light is energy. It is a signal that activates our cells and influences their performance. And it controls your hormones and your metabolism as much as food or even some pharmaceuticals. This means that light has the potential to take years off of your life, or when you use it wisely, to be a performance-enhancing drug.

THE POWER OF LIGHT

Before you discover the most harmful and most beneficial types of light beyond the blue light you read about in chapter 4, it's important to understand how light affects your circadian rhythm. Our bodies and almost all our genes are designed to turn on and off at different times of the day to function in a twenty-four-hour cycle. And our hormones and neurotransmitters naturally rise and fall with this rhythm. To put it very simply, your body releases certain chemicals such as cortisol in response to light that make you feel alert and able to sustain waking activities such as eating, exercising, and so on. Meanwhile, your body releases other chemicals such as melatonin that support sleep, rest, and recovery as a response to darkness.

You read earlier about how clock neurons control our circadian rhythm. There are approximately twenty thousand of these neurons in the base of our brains that act as our bodies' master clock. As our eyes encounter various forms of light throughout the day, they send information to these neurons. About a quarter of these clock neurons can sense *only* blue light. When the eye is exposed to blue light, whether it's from the sun or your iPhone, these neurons send a signal to the rest of the master clock neurons and, collectively, they report to the brain that it is time to wake up and remain alert. Like most animals, we are designed to synchronize our internal clocks with blue light because sunlight is actually the richest source of blue light that exists. We are meant to be highly sensitive to even small amounts of blue light so we can begin the waking process as the sun is just starting to rise in the morning.

But the world we have created is full of light, including artificial sources of blue light that are emitted not just in the morning, but also well after the sun goes down. This is a recent change to our environment. A hundred years ago there was simply no way any human could encounter a light brighter than fire after the sun had set. Now, no matter where we live, we are bombarded with blue light well after dark. As a result, we have disrupted not only our natural sleep cycle, but also the rhythms that keep our bodies functioning optimally.

This disruption has many invisible (and some visible) consequences that influence how we age. There are two steps to halting this disruption. Step one is reducing blue light exposure (so you don't die). And step two is adding beneficial light sources to help you gain Super Human powers . . . just like Superman.

STEP 1: Reduce Junk Light to Live Longer

You read in chapter 4 that excess blue light exposure increases your risk of the Four Killers. One reason artificial blue light is such a problem is because it suppresses your body's production of melatonin, a hormone that tells you when it's time to go to sleep. Low production of melatonin is associated with poor sleep and increases your risk of cancer.[1]

Excess blue light also causes inflammation and mitochondrial dysfunction, primarily because of its impact on glucose control. In the evening, exposure to blue light causes a peak in glucose levels, leading to higher blood sugar and an increase in insulin resistance.[2] This means your blood sugar is higher than it should be, and your body doesn't adequately move that sugar out of your bloodstream. The result is that you are at a greater risk of weight gain[3] and developing type 2 diabetes—which, as you already know, increases your risk of the other three killers.

Yet blue light is not just in your electronic devices. Sadly and problematically, it is everywhere. Most modern light sources contain unhealthy amounts of blue light and far too little of the beneficial spectrums that help balance it out. This is what I call junk light. For example, white LED bulbs lack many of the sun's natural frequencies that our bodies and brains need to function, such as infrared, red, and violet light. They also emit at least five times more blue light than is emitted by sunlight. Fluorescent lights also emit substantially more blue light and less infrared light than sunlight. Shining these junk light sources into your eyes close to bedtime is like Superman taking a little soak in a kryptonite salt bath to prepare for sleep.

In addition to wrecking your sleep and your blood sugar, being exposed to all that blue light, especially at night, taxes your mitochondria by creating excess free radicals in the cells of your eyes. When you are exposed to full-spectrum light, your mitochondria still produce some free radicals, but those free radicals send a signal for the cell to produce extra antioxidants to sop them up. When you are exposed to blue light without the full spectrum of frequencies, the free radicals your mitochondria produce fail to send that signal. Those excess free radicals linger, damaging cells and contributing to all Seven Pillars of Aging.[4]

Exposing yourself to junk light instead of sunlight or incandescent light, which contains less blue light than other artificial sources, is kind of like the difference between eating a normal meal and eating a bowl of sugar containing the same number of calories. Blue light is the high-fructose corn syrup of lighting. And when you consider the fact that the average American spends 93 percent of his or her life under artificial lights, most often fluorescent or LED, it's no surprise that the incidence of all Four Killers is on the rise.

At the same time that we've removed beneficial light frequencies from much of our light sources and replaced them with far too much blue light, we've also begun to completely avoid ultraviolet A (UVA) and ultraviolet B (UVB) light, which are produced by the sun and are necessary for survival. While too much UVA and UVB is damaging, so is too little. When your skin is exposed to UVB light, your body converts vitamin D to its activated, sulfated form. In other words, sunlight makes vitamin D more available so your body can use it.

As we've discussed, having enough vitamin D is essential for longevity. Too little vitamin D plays a role in the buildup of amyloid proteins, one of the Seven Pillars of Aging. Vitamin D is also critical for setting your circadian rhythm and for the regulation of blood sugar. This is one reason why blood sugar levels are naturally lower during the summer months, when you are exposed to more sunlight.[5]

Recent research also has found a direct correlation between a lack of adequate sunlight exposure and the onset of diabetes. One study that tracked a thousand women between the ages of twenty-five and sixty-four for eleven years concluded that the women who had "active sun exposure habits," meaning they were regularly exposed to sunlight, had a 30 percent lower risk of developing diabetes than the women who got very little sun exposure.[6]

By avoiding ultraviolet light completely, we are inadvertently speeding up the aging process. And it's important to note that while I do recommend supplementing with vitamin D, this does not fully replicate the effects of natural sunlight on blood sugar control.[7] Nothing can fully replace the full spectrum of light that we are meant to get from the sun.

In fact, there's a lot of interesting research around the benefits of being exposed to the sun's red and near infrared frequencies *before* you are exposed to UV light. The red and infrared light seems to prepare the cells so they can harness the power of UV rays to produce vitamin D, while naturally protecting themselves from any UV light damage. It also helps the cells recover after UV light exposure.[8] This chronology of light exposure makes a lot of sense, as it corresponds to our circadian rhythm.

Early humans were up at dawn and spent time outdoors in the morning, which allowed them to be exposed to the red and infrared light of sunrise before the sun peaked at midday. Now we spend far more time indoors under unnatural lights. If we do get natural sun exposure, it's often in the middle of the day, when the sun is at its most intense. Our cells don't have a chance to prepare for that UV exposure without the help of red or infrared light. Is this why incidents of skin cancer have increased despite our frequent and often obsessive use of sunscreen? I wouldn't be surprised if lack of exposure to these frequencies plays a role in the severity of the sun damage we experience now.

It's ironic that we avoid the sun to protect our health, while in reality adequate but not excessive sun exposure helps us live longer. A recent study of twenty-nine thousand women in Sweden who were tracked over twenty years concluded that "avoidance of sun exposure is a risk factor for death of a similar magnitude as smoking." The study showed that people who avoided the sun had a reduced life expectancy of between 0.6 and 2.1 years.[9]

This may sound surprising, but the circadian rhythm, which is controlled by light, is fundamental to all life's survival on this planet. Animals, humans, plants, and even fungi all go to sleep and wake up in twenty-four-hour rhythms. This is so ingrained that if we took an animal or plant from our planet and put it on Mars (or any other planet that has a day/night cycle that's not twenty-four hours), it would struggle to adapt and might not survive. That is profound.

Dr. Satchin Panda from the Salk Institute has studied the impact of the disruption of circadian rhythm on health and longevity. He says that when animals in a laboratory have a mutation in their circadian clock, their risk factors increase for a number of killers, including diabetes, obesity, cardiovascular disease, and cancer. In fact, when humans' circadian rhythms are altered in a lab (study participants are restricted to five hours of sleep), they begin to exhibit symptoms of these killers within only a few weeks.

Listen, I'm not suggesting that you stop using electricity or move out to the middle of the woods somewhere or live naked in the sun. There are ways to enjoy the conveniences of modern technology without harming ourselves in the process. In fact, you can use natural and beneficial sources of light to nurture your circadian rhythm and prevent and even reverse much of the aging and disease that junk light can cause.

One simple way to do this is to install red lights, which include more of the sun's natural frequencies. Red light is at one end of the visible light spectrum, the electromagnetic spectrum of light that's visible to the human eye. Meanwhile, infrared light has a wavelength that's just beyond the red end of the spectrum. You can't see infrared light, but you can feel it as heat. It is the reason why exposure to direct sunlight makes you feel warm.

At my house on Vancouver Island, all the exterior lights are red. My friends may tease me for giving the appearance of living in a submarine or some kind of house of ill repute, but I'm able to go outside at night without disrupting my circadian rhythm. Plus, I can see the stars. In addition to benefiting my health (and the health of my family), red outdoor lights are also healthier for the animals living near my home. Unlike white LED porch lights, my red lights don't attract bugs. This matters more than you might think, especially since most people who attract bugs with junk light go on to kill them with pesticides. According to researchers, 40 percent of the world's insect populations are in decline and could die out in the coming decades. You know, the years you're planning to live through. The lead author of this paper from the University of Sydney in Australia says, "If we don't stop it, entire ecosystems will collapse due to starvation."[10]

The other reason I use exterior red lights is that it seems irresponsible to knowingly mess with the circadian rhythms of animals around my home. I have three different species of owls that nest within a hundred yards of the house because our red lighting doesn't ruin their habitat. It wasn't until I saw this that I realized how much junk light affects every living thing on the planet. So get some red LED bulbs for your landscaping, and get ready to field questions from your neighbors. They'll either think you're cool or proposition you. Either way, you'll live longer.

Another way to use light to your advantage is to install light sources inside that give off the full range of frequencies. At the Bulletproof Coffee shops we have a big light box around the door that changes color based on what time of day it is to help promote a healthy circadian rhythm. And at our new Bulletproof headquarters we have low voltage halogen lighting on dimmers and backup LED lighting so we can add red and amber frequencies in the evening to mirror the natural rhythm of the sun.

One of the easiest ways to make the light in any environment healthier is simply to install dimmers that allow you to dim the main light source and change its richness. If you have an opportunity to switch out the lights in your home or office environment, I strongly encourage you to take color spectrum into account. Don't make yourself crazy about getting it perfect all the time—stress like that makes you old. Just aim to get it mostly right at home, and you'll be so far ahead of the curve that you'll feel it.

When it comes to blue light, we've already discussed how to reduce your exposure from electronic devices in chapter 4. Here are some additional ways to eliminate aging light sources from your environment:

- Starting at 8:00 P.M., dim all the lights in your home. Switch to lighting your environment with red LEDs (or candles if you want extra credit).

- If you can't install dimmers and/or have to be in a brightly lit environment after sundown, wear glasses that filter out blue light. If you've seen me onstage, you've undoubtedly noticed the yellow-tinted glasses I wear indoors to filter out some (but not all) blue light. I probably would have gotten picked on in high school for this, but now I am convinced that they look cool.

- If you don't want to use glasses that filter out blue light but have to be in a brightly lit environment after dark, wear sunglasses if appropriate.

- Go outside for ten to twenty minutes a day to get adequate (but not too much) UV light exposure. Try to do this in the morning before the sun is at its peak, and you won't need sunscreen.

STEP 2: Add Beneficial Light Sources to Look Better and Have More Energy

Now you know how to avoid the junk light that makes you old. But what if you could also selectively *add* frequencies of light to your environment to help your body not only survive but thrive? Just as the wrong light frequencies overtax your mitochondria and cause inflammation and aging, certain beneficial light frequencies can improve mitochondrial function.[11] This reduces inflammation[12] and helps your body produce energy more efficiently, which leads to fewer hits to your system and ultimately a younger and more kick-ass version of you.

Let's take a closer look at the light sources that can help make you Super Human.

RED/INFRARED LIGHT

About twenty years ago, I was in two similar car accidents that each caused whiplash. After the first accident, it took me nearly a year of pain and headaches using all kinds of therapies to fully recover. The second time, right after a large BMW plowed into the back of my unmoving Ford at a stoplight, my first thought wasn't about my car. It was dreading another year of suffering. Sure enough, within two days, my neck was sore, my right arm was cold and numb, and I had a dull headache. But by then I was keyed into the anti-aging community, so I had access to experts who could help me heal. A friend introduced me to a naturopath who met me for dinner in San Jose, California. In the parking lot after dinner, he pulled out a small, handheld medical laser and told me to put it on my upper back where I was sore. The laser was meant for racehorses and wasn't even approved for use on humans.

I was dubious at first, but when I realized it wasn't the kind of medical laser used for burning or cutting, I gave it a try. Within just three minutes of using this laser to pulse red and infrared light onto my upper back, I felt a bolt of electricity shoot up my spine as the knotted muscles released. My arm grew warm, and my head cleared. The relief was faster than anything I'd ever experienced.

I asked how much the device cost, only to find it was 50 percent of what I made in a month. When I compared that to spending a year going to physical therapy, chiropractors, and massage therapists while in pain, the medical laser looked like a bargain. I bought it, and I've been pursuing light therapies for the two decades since then.

Red/infrared light therapy uses these wavelengths of light together to restore, repair, and protect tissue that is injured, degenerating, or even at risk of dying by activating the stem cells that normally become latent as we age while improving mitochondrial function.[13] This actually increases the amount of energy your mitochondria can produce.[14] Red/infrared light therapy also increases levels of nitric oxide, an important molecule your body produces that keeps blood vessels healthy. Having more nitric oxide increases circulation, which ensures that all of your cells are nourished by blood, oxygen, and nutrients.[15]

Infrared light therapy also helps create exclusion zone (EZ) water, the type of water inside your cells. When you don't have enough EZ water, your cells become dehydrated and their mitochondria stop functioning well. You can get more EZ water by drinking raw vegetable juices, fresh springwater, or glacial meltwater. And infrared light exposure causes EZ water to spontaneously form in your cells, boosting mitochondrial function. My company donated $50,000 to the University of Washington to study this recently discovered form of water that is important for our biology.

Infrared light therapy is also extremely effective at treating muscle fatigue and injuries,[16] factors that keep too many people from getting adequate exercise as they age. Think about that—this one technology gives you advantages in several pillars of aging. More stem cells mean you can replace old cells, more mitochondrial function means less inflammation, and an increase in exercise and circulation mean you can prevent cell loss!

While each wavelength affects your body differently, the most effective wavelengths for healing range from 630 nm to 670 nm (red) and 810 nm to 880 nm (infrared). Light at these wavelengths penetrates up to 8 to 10 millimeters into the skin and affects your cells on a biochemical level. Red light deployed very close to the skin can impact all the surrounding skin layers, blood vessels, lymph pathways, nerves, and even hair follicles. This decreases inflammation, improves skin tone, repairs sun damage, fades scars and stretch marks, promotes hair growth, and even stimulates collagen production in the skin, which reduces the appearance of wrinkles and prevents new ones from forming. It also heals wounds[17] and age-related macular degeneration of the eyes.[18]

There are many ways to take advantage of red and near infrared light therapy. Many qualified professionals offer these treatments, including dermatologists and estheticians, trainers, and sports medicine practitioners. Some medical spas offer red light therapy, and we offer it at Upgrade Labs, the chain of human upgrade facilities I started. You can reap the health benefits of red and near infrared light therapy in your home with personal red light therapy devices like the medical-grade options from Joovv. Somewhat similar to LEGO blocks, their unique modular design allows you to build out a full-body system over time. You can also pick up some inexpensive red and infrared bulbs for $25 to cut down on artificial blue light exposure at night. Keep in mind that simply installing red lights around your home like I do helps your circadian rhythm by balancing out excess junk light, but it is not light therapy. You have to use bright LEDs placed very close to the skin to be effective.

Being a professional guinea pig has its perks, such as having a full-blown biohacking laboratory at my home on Vancouver Island. I use a medical-grade red light therapy machine that looks like a tanning bed. Instead of frying my body with harmful waves, its more than forty thousand red and infrared LED lights illuminate my entire body to help me recharge my mitochondria and grow healthier collagen. The difference in my skin since I started using it is remarkable.

Of course, this kind of thing is generally cost prohibitive to put in your home (a Tesla is a better investment), plus it takes up space—so I recommend going to a facility to use a bed like this for forty minutes starting at once a week. You can also get results with inexpensive red light panels you install at home; the only drawback to these is that they cover less of the body. Still, using smaller red light panels at home has proved to be a surprisingly effective treatment for my occasional nausea and headaches, and I've lost track of how many times my wife, Dr. Lana, and I have used these panels to successfully treat our kids' stomachaches and other pains.

Another way to benefit from infrared light is to spend some time in an infrared sauna. These differ from regular saunas in that they heat the body from the inside out rather than from the outside in. While traditional saunas heat the air around you, in an infrared sauna the light directly penetrates and heats your body tissue. As a result, infrared saunas don't get as hot as traditional saunas. This means you can stay in these saunas longer without feeling like you're going to pass out. I've been using an infrared sauna for years, initially to detox from mold and mercury exposure, and I've found it to be hugely beneficial.[19]

Spending time in a sauna (infrared or not) also leads your body to produce heat shock proteins (HSPs), which prevent protein degradation caused by oxidative stress. HSPs scavenge free radicals, boost glutathione levels, and make sure proteins retain their proper structure, protecting you from several pillars of aging.[20] You can find infrared saunas at day spas, yoga studios, and gyms. Start with twenty- to thirty-minute sessions two or three times a week, and work your way up from there.

YELLOW LIGHT

Also known as orange light or amber light therapy, yellow light therapy utilizes wavelengths in the range of 570 nm to 620 nm. Yellow light therapy increases your mitochondria's ability to produce ATP, cellular energy, but it doesn't penetrate as deeply as red or infrared light. It is mostly used to treat skin conditions such as spider veins, rosacea, and damage from too much sun exposure.

Yellow light therapy is relatively unknown in the world of biohacking, but many studies have demonstrated the effectiveness of yellow light therapy on the skin. In a large study that tracked nine hundred patients over two years,[21] participants received either yellow light therapy alone or in combination with other light therapies. The patients who received yellow light therapy alone saw a softening of the skin, a reduction in fine lines, and a decrease in photoaging (aka sun damage). Another study revealed a reduction in signs of photoaging in 90 percent of subjects using yellow light therapy. Their skin had a smoother texture and a reduction in hyperpigmentation (age spots).

Most notably, 100 percent of the patients showed a marked increase in collagen production.[22]

Additional studies show that yellow light therapy speeds healing after a patient receives other types of more invasive therapies such as laser skin resurfacing.[23] Given these findings, researchers have begun exploring yellow light therapy to help patients with cancer recover more quickly from radiation treatments. In one study, a group of patients undergoing radiation also received yellow light therapy, while a control group only received radiation. Of the control group, 68 percent had a painful skin reaction from the radiation, while only 5 percent of the group receiving yellow light therapy had a similar reaction.[24] That's huge!

The only multi-waveform amber light device on the market as of the time of this writing is the TrueLight energy square, which emits red, infrared, and amber lights and costs a few hundred dollars.[25] However, it is not approved for any medical uses. Doubtlessly many more devices will soon become available.

We'll discuss some additional options for reversing the visible signs of aging later on in chapter 13. For now, though, remember that it's much more effective to prevent and reverse your aging from the inside out and then use strategic anti-aging tools to correct any remaining damage on the outside. Treating the visible signs of aging first will take longer, be less effective, and do nothing to prevent the cellular damage that caused the aging to begin with. When your mom told you, "It's what's on the inside that counts," she wasn't kidding.

INTRAVENOUS LASER THERAPY (IVL)

If you really want to go all in with light therapy, you can try intravenous laser therapy (IVL), which was developed in the former Soviet Union in the early 1980s. This is a type of oxidative therapy with hormetic effects, meaning that, like exercise, it stresses your cells a strategic amount, causing them to strengthen in response. A doctor administers IVL by introducing UV light directly into your bloodstream. In studies, it has been shown to reduce inflammation, boost mitochondrial efficiency, and stimulate the production of more ATP.[26] It also produces a huge wave of activated vitamin D_3, far more than sun exposure can achieve. Though Russian scientists have been using IVL to treat heart disease and many other conditions since the 1980s, odds are you've never heard of it because the American medical establishment hasn't yet embraced this therapy.

Recent anti-aging research from Dubai shows that by replenishing blood cells, IVL affects the vascular system, the immune system, and the endocrine system. The study concluded, "It can lead to lower the incidence and number of vascular diseases, and indirectly to the reduction of the number of diseases in other organs and even systemically, thus helping to prolong the lifespan."[27] I aim to do IVL once or twice a year. It costs about $250 for a treatment. Unfortunately, it is still hard to find practitioners who administer IVL outside of a few major cities, but it's remarkably cost-effective.

For most people, the interventions you've learned so far—eating the right food, getting enough quality sleep, avoiding junk light, and adding exposure to beneficial light—add up to enough of a defense strategy to avoid or lesson many of the cuts that cause aging. I wish I'd had access to this knowledge when I was suffering as a young person. Making simple yet consistent changes to my diet, sleep, and light exposure has helped restore my function as a regular human.

Seeing the profound impact that these relatively small changes could make, I was inspired to keep going, to continue undoing the damage I had unwittingly caused and seeing if I could proactively prevent taking so many hits in the future. If I could feel and perform so much better after changing my environment, sleep, and dietary routines, what would become possible for me if I took things up a notch?

In other words, I had successfully stopped dying. Now it was time to start aging backward.

Bottom Line

Want to not die? Do these things right now:

- Make sure you are exposed to some red or infrared light every day, or aim for fifteen to twenty minutes of natural sun exposure a day.
- Protect yourself from junk light by changing the bulbs in your home, and get some glasses that block some (but not all) blue light. Don't be afraid to wear them indoors after dark—even in public! Sunglasses will help a bit if you don't want to purchase special ones, but they're not as effective.
- Consider trying an infrared sauna to aid in detoxification and boost your mitochondrial function. For help with wound healing, muscle fatigue, or tissue repair, look into red and infrared light therapy. And if your concerns are primarily skin-deep, yellow light therapy may be an easy fix.

PART II

AGE BACKWARD

If you stop reading this book right now and just focus on not dying, you're already destined to hit the Super Human level. The stuff you've already read about that is going to kill most people also makes them weaker, slower, and less happy with age. If you just hold the line on the Four Killers, you're already going to perform better than most of your peers by the time you're old. Unless, that is, you cultivate friends who share in this knowledge and put it to work, in which case you'll obviously have Super Friends. (Sorry, couldn't resist it.)

The problem is that if you're anything like me, you've already taken more hits than you'd prefer. If aging is death by a thousand cuts, or at least death by seven decomposing pillars, wouldn't it be great if you could reverse some of the damage you've sustained so your biology would actually get younger? This is a new kind of Super Human power: to *reverse* the effects of aging. When you start by doing things that will help you avoid dying and then work on aging backward, you'll experience a better life on every level.

Family and friends were amazed to see how drastically I had transformed my biology and halted my premature aging by making relatively small changes to the environment around me. What they didn't know is that along the way I was fortunate to tap into a powerful community of top scientists and researchers who were exploring anti-aging technologies unheard of in mainstream science. This provided the opportunity to keep pushing further so that I could surpass the typical cosmetic anti-aging approaches and actually begin to age backward from the inside out.

Some of the things I've tried since then were new and by definition unproven, but all of them met the risk-reward ratio I wanted, and all of them had enough evidence to be plausible. I'm not here to encourage you to try everything I've done. In fact, you probably shouldn't, even if you also are setting out to overcome a subpar start in life and you share my risk-reward perspective. Do your research and make decisions based on *your* body and *your* goals, not mine, and work with professionals on things that are higher risk.

Even if you never try the advanced techniques, there are plenty of simple things you can do to get more energy and more power in the short term and continue aging backward in the long term. After all, if you start now so you live longer, there will be plenty of time for "unproven" techniques to become more affordable and mainstream. I'd feel bad if you missed out on all the amazing anti-aging things that are coming your way because you didn't get started on "not dying" right away.

It's totally okay to delay the expensive stuff—it will get cheaper eventually. For perspective: The cost of sequencing the first human genome was $100 million in 2003. As I write this, you can get your entire genome sequenced for free if you share your data, or you can pay a few hundred dollars and get your data in a week or two. Rejuvenation technologies are on the same path.

TURN YOUR BRAIN BACK ON

As your cells die off with age and are not replaced (thanks to the Seven Pillars of Aging), your brain begins to shrink and, over time, degenerate. This is why, when we picture an old person, we see that dreaded image of someone who can't remember his or her own name. Well, I believe we've arrived at a point where we can and should erase that image for good and replace it with a new one—an image of someone who is as mentally sharp at eighty as he or she was at twenty. Because the truth is, if you take care of your brain, your brain will take care of you. And even if you are already experiencing some signs of brain aging, you can stop them and reverse the damage starting today. I know this because I've done it.

When I got the results of my first brain SPECT scan, it provided undeniable proof that the cognitive dysfunction I'd been experiencing for years was not only real but also had a physical basis. It was one thing to have a doughy, stretch-mark-covered, aching body, but a prematurely aging brain was quite another. In addition to my poor performance on exams while in business school, every day I experienced many of those tip-of-the-tongue moments when the right word just wouldn't come to mind, and I'd often lose track of what I was doing in the midst of some ordinary task. I'd find myself at the store not knowing what I wanted to buy or staring at my to-do list trying to remember what tasks I'd already decided to tackle.

Over the years, I had grown accustomed to being fat, often in pain, and unable to perform certain physical activities. But my one saving grace had always been that I could think. I may have been obese and socially awkward, but at least I was smart. Developing the equivalent of an aging brain threatened my entire identity, not to mention my sense of security. After all, I relied on my brain to earn a living and put a roof over my head. So the prospect of being unable to perform cognitively was terrifying on a profound level.

This was a turning point. Knowing that if I didn't figure out how to fix my brain, I wouldn't have the future I desired, I was determined to do whatever it took to repair it.

While I was clearly far too young to be experiencing symptoms of cognitive dysfunction in my twenties, the truth is that mental decline shouldn't be considered "normal" at any age. There is no good reason why any of us need to experience impaired cognition as we get older, but it is so common that we jokingly refer to instances of forgetfulness as "senior moments" instead of calling them out for what they are—symptoms of dysfunction that are also precursors of Alzheimer's disease and senile dementia.

Even if you are still young and aren't yet experiencing any cognitive problems, it's a good idea to take action to prevent brain degeneration before it starts. As a bonus, doing so will also improve your cognitive performance right now. While you may not feel the effects of lengthened telomeres immediately, you will notice right away when you have more brain energy. It's the Bulletproof feeling of being able to really bring it without trying too hard. Your performance can be as effortless as it is effective.

You probably won't start experiencing "senior moments" until you're forty or fifty (or maybe just really hungover), but it's possible to feel the effects of cognitive dysfunction much earlier. When I interviewed Dale Bredesen, MD, an internationally recognized expert in neurodegenerative diseases and the author of *The End of Alzheimer's*, he told me that people are typically diagnosed with Alzheimer's a full twenty years after the underlying pathophysiological process begins. Since some people are diagnosed at forty, this means the physical problems behind this killer can start as early as age twenty. It also means senior moments in your forties may be an indication that you'll be diagnosed with Alzheimer's in your sixties. The Four Killers take time to do their work.

This isn't frightening—it's a weakness you can exploit to beat Alzheimer's. You have two advantages: The development of Alzheimer's is slow, and we're getting better at seeing it earlier. There's a lot you can do to improve your brain function right now and for the long term. Who knows what my brain function would be like now if I hadn't intervened twenty years ago? We've been led to believe that whether we get Alzheimer's or senile dementia is up to either genetics or the luck of the draw, but that's just not true. The same interventions I used to reverse my cognitive dysfunction can reduce your risk of developing Alzheimer's and help you think and perform better at any age.

NEUROFEEDBACK FOR DUMMIES

Long before I got worried enough to get a SPECT scan, I could feel that my brain didn't work the way it should. It was a nebulous, hard-to-put-my-finger-on feeling, but I knew that a spark was missing. I felt slower than I should have been. Luckily, some of my friends in the anti-aging community knew about a new modality of brain treatment called neurofeedback that was almost completely unheard of back then. They helped me find radically effective practitioners and treatments that I otherwise never would have known about or had access to. One of the people they introduced me to was a Bay Area chiropractor who also performed neurofeedback treatments.

I had no idea what to expect at the first appointment. Pictures ran through my mind of a big facility with blinking lights, men in white lab coats, and maybe something from the set of *Tron*. When I arrived, though, I found a humble office with a fish tank and a beaded curtain separating the lone treatment room from the waiting area. In the waiting room, a young boy whom I did not know was autistic walked up to me and started screaming at the top of his lungs while he ran in circles around me. It was not what I was expecting from my first neurofeedback experience.

When I was called into the treatment room, I sat in a converted closet as the doctor glued two small electrodes to my scalp using sticky white paste that didn't wash off easily that night. He was preparing me for a test called an electroencephalogram, or EEG. At the time, doctors used it almost exclusively to diagnose epilepsy, sleep disorders, and other brain abnormalities by measuring electricity coming from the brain without actually getting an image of the brain itself. I would not be looking at a picture of my brain to see what was wrong with it; instead, the idea was to show my brain (and me) what it was doing with real-time feedback. I was excited to become awesome at self-regulating.

While wearing the sensors, I played a primitive video game and watched the electrical patterns of my brain on a screen. Every time the phone rang in the office, I could see the EEG going crazy. "Do you see that spike?" the doctor asked. "That's your fight-or-flight response. You go into panic mode when a phone rings."

The fight-or-flight response is a hardwired survival mechanism that activates the sympathetic nervous system, causing your body to release stress hormones such as cortisol so you can run or fight and to send blood flow away from the prefrontal cortex, the part of the brain involved in high-level decision-making. It makes sense that a loud noise such as a phone would activate the fight-or-flight response. The sympathetic nervous system can't distinguish between an innocuous sudden sound like the ringing of a phone and a real threat like a nearby explosion. In the case of an actual threat, you don't want a lot of activity in the prefrontal cortex, because this would lead to overanalyzing instead of acting. To survive an attack, you want to start running instead of standing still while you weigh the pros and cons of running.

But you *do* want to be equipped to weigh the pros and cons when it's just a phone call (or an exam, or even a stressful email, for that matter) activating your fight-or-flight stress response because you haven't learned to control it! And nothing will make you older faster than out-of-control stress. My fight-or-flight response was being activated far too often, which is not uncommon in today's stressful world. And it's pretty hard to hone your concentration or improve your decision-making skills if your brain is constantly and easily panicked. Neurofeedback can help you learn to self-regulate so your fight-or-flight response isn't activated quite so easily.

Over the next several weeks, I returned for regular neurofeedback treatments and worked to rewire that stress response so my brain no longer felt like it was constantly under threat. Neurofeedback appears passive, but it is actually hard work. Your brain responds to visual and auditory input by changing its electrical output as it learns to self-regulate. I could feel some subtle changes and see on the EEG output that the sound of a phone ringing no longer caused as large a spike.

When I went in for my sixth treatment, the same young autistic boy was in the waiting room. This time he walked up to me, looked me in the eye, and said, "Hi, my name is Bobby." I was floored. He exhibited such a dramatic transformation that it still stands out to me twenty years later. I thought to myself, *I must have this technology at home! What superpowers will I be able to develop when I understand everything in my brain?*

As a result, I have owned a collection of EEG machines since 1997. Neurofeedback technology is now much more effective and affordable, yet for advanced work there's nothing like working with a trained clinician to help you gain control of your brain—or even upgrade it. Neurofeedback is also commonly and effectively used to help patients heal from trauma and symptoms of anxiety and depression. It is now available in most major cities at clinics that charge about as much as a massage for an hour of neurofeedback, and there are even home units available.

My encounter with Bobby proved to me that our brains are capable of profound change. I'm walking proof of it, too. Despite having had serious brain problems, I now have a hippocampal volume that is in the 87th percentile for someone my age. As you read earlier, that part of the brain typically shrinks as you age, so I have at least held the line on that form of brain aging and have likely improved on where I was in my twenties. I have also increased my IQ and even my working memory.

But Bobby and I are not unique. We are all capable of forming new neurons and new connections between neurons at any age so your brain can continue learning and growing. This is called neuroplasticity. A scientist named Eric Kandel won the Nobel Prize in 2000 when he proved that the brain is capable of reorganizing itself by forming new neural pathways throughout a lifetime. (Previously it was believed that the brain was not capable of doing this past your mid-twenties.)

After all this work, my resilience is higher than it's ever been, my stress response is low, and I almost never have to search for a word that won't come to mind. It's so unusual that when I do, I can easily trace the reduced brain performance back to something in my environment or a food I ate that caused inflammation and impacted my brain function. In other words, it's not age that causes these things to happen to us—it's the accumulation of cuts that keep our mitochondria from working at full throttle.

To fully understand the connection between inflammation and brain dysfunction, you need to understand the role of microglial cells, which you read earlier are the immune cells of the brain. Neurons get all the press, but they don't work without the microglia. Previously you read about amyloid plaques that build up in the brain when misshapen proteins clump together. The microglial cells are supposed to help clear out those amyloids when they develop in the brain. But when excess amyloid plaques develop, microglial cells accumulate around them, releasing more and more inflammatory compounds in an attempt to clear them away.[1] These microglial cells are trying to heal the brain, just as immune cells create inflammation to heal the body when it's been injured. But when this becomes a chronic condition in the brain, it damages neurons and leads to neurodegenerative disease.

Chronic inflammation in the brain from bad food, chronic stress, heavy metal or toxic mold exposure, infections, or circadian disruption also leads your microglial cells to create too much of a protein called progranulin (PGRN).[2] High PGRN is associated with Alzheimer's, Parkinson's, ALS (Lou Gehrig's disease), tumors, and all kinds of bad stuff that will keep you from ever becoming Super Human.

The good news is that you can avoid many of the cuts that cause inflammation in the first place. According to Bredesen, the biggest risk factors for developing Alzheimer's disease are chronic inflammation, insulin resistance, and exposure to toxins. These are all environmental factors, not genetic ones! And it's much easier to hack your environment to avoid Alzheimer's than it is to try and reverse it. With the techniques I've honed over the last twenty years, I have learned to experience far fewer cuts and successfully reverse my cognitive dysfunction. And when I do take a hit, I know how to counter it so my brain is back up and running at full speed as quickly as possible. The resulting resilience is something I am counting on as I work to live beyond a hundred and eighty.

In addition to neurofeedback, I've found that the most effective interventions for preventing and reversing cognitive dysfunction fall into three main categories: light, food, and drugs. Since we most recently discussed light in the previous chapter, let's start there.

SHINING A LIGHT ON THE BRAIN

Even today, most conventional doctors will not recommend pointing a laser at your brain, but those in the anti-aging community have been doing just that with great success for years. After using laser therapy on my whiplash back in the 1990s and experiencing such profound effects, I began researching other ways to use light as an anti-aging tool.

Deep in the corner of the young Internet I discovered a man who had experienced cognitive dysfunction of his own and found a way to hack it. He started a Yahoo group to share his discovery. At the time, there was no research supporting the safety or efficacy of using infrared light on human brains, but there were studies showing it increased blood flow in animals. That was enough for this guy to create his own high-powered infrared LED device, which looked like someone had hand-soldered an LED and stuffed it in a pill bottle with a hole in the lid, because that's basically what it was. He sold a couple hundred of them through the Yahoo group and wrote about his experience shining the invisible LED light down the middle of his brain for two minutes a day and dramatically improving his brain function, focus, and mood. For him, the risk-reward equation made it worth it to test the laser.

I liked the idea, so I bought one of his devices and used it for two minutes every day on my forehead and the back of my head. It literally felt like someone had flipped a switch in my brain and turned it on. Everything was easier. I was used to pushing so hard all the time, but now I felt as if I was being pulled. That device became one of my most treasured possessions because it made my brain work better every time I used it. I still have it in my closet, even though today I have much better brain LED devices in my labs.

Now I understand that the infrared light from the LED restored mitochondrial function when I focused it on parts of my brain, creating more energy that I was able to use to think clearly for the first time in ages. Today there is plenty of research to show that infrared LEDs and lasers can effectively treat neurodegeneration by stimulating mitochondrial function in the brain, therefore increasing cellular energy production.[3] Lasers can even directly benefit neurons that have been functionally inactivated by toxins.[4] This may very well be why laser therapy was so helpful in healing my brain from all those years of toxic mold exposure.

Since infrared light can penetrate through muscle and even bone tissue to reach the brain, it's quite effective at providing a number of brain function benefits. In the past few years, doctors have used infrared lasers to treat strokes, traumatic brain injuries, degenerative brain disease, spinal cord injuries, and peripheral nerve degeneration.[5] And long before you develop symptoms of those degenerative diseases, it can enhance brain function in normal, healthy people.[6]

But only twenty years ago, it was considered nearly pathological to even consider shooting a laser at your brain. That didn't mean it was any less effective, though. Remember the guy who created the device I used? About two years after he started the Yahoo group, he wrote that his brain was now working so well that he was going to go to medical school. A week later, he deleted the group and all its postings, likely because he was concerned it could lead to liability issues after he became a doctor. There isn't a scrap of information about it online anymore. It's ironic to fear talking publicly about the very tool he used to get through medical school. I have no idea where this guy is now, but I hope he's somewhere helping people turn their brains back on. He definitely helped me.

Before you grab a laser and start shining a light on your brain, please keep in mind that light therapy is incredibly powerful. If used incorrectly, it can hurt you. I once fell asleep with an infrared light shining on my leg and woke up with a second-degree burn that took six weeks to heal. But much scarier was the time I decided to try my early LED brain device to hack the language-processing part of my brain.

It always bugged me that I suck at hearing the nuances of other languages, especially after I married Dr. Lana, who speaks five languages fluently. So I used the device on the left side of my head over the brain's language-processing center for two minutes. For the next few hours my speech was completely garbled. No matter how hard I tried, I couldn't talk normally. I was scared shitless (that's the technical term) until my normal speech was restored, because I made my living in part onstage teaching audiences about the future of technology. That is the risk part of the risk-reward equation. By going slowly, I avoided long-term problems. Who knows what would have happened if I'd stimulated my brain for an hour?

We know a lot more now about how to safely use these therapies. Many years after I became an early adopter of laser therapy for the brain, I attended the Near Future Summit, an event focusing on innovation that's frequented by the TED crowd in San Diego. They wanted to mix it up that year, so they hosted a pajama party for big-name venture capitalists and other conference attendees. Apparently, unicorn onesies were in vogue with VCs at the time. Meanwhile, not knowing about the party ahead of time, I asked my assistant to order something on Amazon for me to wear. She got a full set of red silk Hugh Hefner–style pajamas. Great.

Somewhat embarrassed, I sat down next to a neuroscientist from MIT who had neglected to wear pajamas in favor of regular clothes. She had given a speech earlier that day about using a very simple light therapy to reverse Alzheimer's disease. In her research, she discovered that lights that blink forty times a second can break up amyloid tangles in the brain. Her goal was to install flashing light panels in all nursing homes, something I expect I'll live long enough to witness.

It's only now that are we beginning to see well-funded research for these technologies. In 2016, MIT researchers demonstrated that LED lights flickering at a specific frequency could substantially reduce the amount of beta-amyloid plaques in the visual cortex of mice.[7] Not only did this treatment cause the mice to produce fewer amyloids, but it also invigorated microglial cells, the immune cells in the brain that are normally responsible for destroying amyloid plaques.

The next year, a study out of Harvard Medical School and Boston University School of Medicine showed that patients with dementia experienced significant improvements in cognition when treated with laser therapy. For this small human study, five patients with mild to moderately severe cognitive impairment received twelve weeks of transcranial laser treatment. After the twelve weeks, they saw a significant improvement in brain function, slept better, experienced fewer angry outbursts, and felt less anxiety with no negative side effects.[8]

There is an affordable laser treatment device that can be used at home with a headset that emits near infrared light through diodes placed on the scalp and inside the nostril called Vielight. They are conducting a clinical trial with 228 participants across North America to see what it does for Alzheimer's. If someone I love was suffering from Alzheimer's right now, I wouldn't want to have to wait for this trial to be over before getting the device. The risk of allowing Alzheimer's to progress is much higher than the risk of trying it out. Devices that use light on the brain range from $200 to many thousands of dollars.

FEED YOUR BRAIN

Anything you eat that causes inflammation is going to hurt your brain function—period, end of story. But there's more you can do to protect your brain than just avoiding inflammatory foods. As you get older, it's important to eat a diet that consistently keeps your blood sugar low, avoids spikes, and keeps ketones present in your blood. In the last ten years, studies have consistently demonstrated that insulin resistance is at least partially responsible for amyloid plaque formation in the brain.[9] As you read earlier, the latest research connecting brain degeneration with insulin resistance has led many experts to begin referring to Alzheimer's as type 3 diabetes.

Remember, insulin's job is to lower blood sugar levels by moving sugar out of the bloodstream and into your cells, where your mitochondria burn it for fuel. If you eat too much sugar, your body produces more and more insulin to move it all out of the bloodstream, but it has nowhere to go because your mitochondria can't burn it fast enough. This is the beginning of insulin resistance, which you know is a major precursor to type 2 diabetes. That high blood sugar also leads to the creation of AGEs and amyloid plaques, which are two of the Seven Pillars of Aging.

In order to ensure that your blood glucose levels don't drop too low when you have an abundance of insulin, your body produces insulin-degrading enzyme (IDE) to break down excess insulin. Interestingly, IDE also helps destroy the amyloid plaques that cause Alzheimer's in the brain and aging throughout the rest of the body. But it cannot break down excess insulin and destroy amyloid plaques at the same time. If IDE is constantly busy breaking down insulin, there is not enough left to fight your amyloid plaques, creating the opportunity for them to build up in the brain.

So when you eat a lot of foods that spike blood sugar, your body produces tons of insulin, and your IDE has to constantly break down insulin to get that sugar out of your bloodstream. This leaves the gates wide open for amyloid plaques to develop and for you to age rapidly and potentially develop Alzheimer's. This means that one of the easiest and most effective ways of reducing your risk of Alzheimer's is to simply stop eating sugar. This way, your IDE can focus on breaking down amyloid plaques instead of constantly working to break down excess insulin.

A powerful intervention is to take 400 to 1,000 mcg of chromium picolinate daily with 25 to 100 mg of vanadyl sulfate, ideally at the same time you eat carbohydrates. These minerals lower the blood sugar spike that occurs after meals, even if you have healthy blood sugar levels. In diabetic animals, vanadyl sulfate lowers blood glucose, cholesterol, and triglyceride levels.[10] And chromium reduces glucose levels and insulin resistance to help prevent type 2 diabetes.[11] These supplements are quite affordable, but evidence suggests taking higher doses than the government recommends.

You can also make sure that when you do consume carbohydrates, especially sugar, you pair them with foods that contain plenty of fiber or even with some saturated fat because this helps you avoid a blood sugar spike. To be clear, the combination of sugar and fat is bad for you, but sugar alone is even worse. For instance, ice cream raises blood sugar less than drinking a soda that contains the same amount of sugar.

In 1998, as I struggled with low energy and obesity, my doctor looked at my blood tests and said, "Maybe you have high blood sugar." The next day I bought a $200 blood sugar monitoring device at the drugstore; it enabled me to prick my finger to get a reading. When

I returned to my doctor's office two weeks later with sore fingertips and two pages of data from dozens of tests per day, it was pretty clear that my blood sugar was a little high but not high enough to cause my symptoms. The doctor thought I was crazy. He said, "Those blood glucose meters are for diabetics. You're not diabetic [yet], so you shouldn't use it." I may have been crazy, but I learned these techniques to avoid big blood sugar spikes by monitoring my blood sugar.

Today you can buy a blood glucose meter for about $20 at a drugstore; the meter enables you to prick your finger and know what your blood sugar is doing. If you're really into living a long time, you can invest more by purchasing a twenty-four-hour continuous blood glucose monitoring system like I use. It's a painless coin-sized device that sticks to your triceps for up to two weeks and shows you your blood sugar level on your phone at any time. You can see what your blood sugar does when you sleep, after exercise, and after meals.

The first time I tried this device, I stuck it on my arm and hopped on a plane to New York to be on Dr. Oz's show. I was wearing my Oura sleep tracking ring and the glucose monitor on the same arm, and when I got to the TV set, the producer asked if I could take off the weird metal dot on my triceps. When I explained what it was, she said, "It will look interesting on camera—like you have a robot arm." My brain is worth it.

Reducing blood sugar is the easiest step to reducing your risk of Alzheimer's, but cyclical ketosis is even more powerful. This allows your body to alternate between fat and glucose as fuel for maximum resilience. This metabolic flexibility is important for brain cells that will otherwise become insulin resistant and unable to burn glucose efficiently, and it provides your neural mitochondria with their preferred source of fuel. In my personal quest to live to a hundred and eighty, I make sure to always have some ketones present in my bloodstream, even when I do eat carbs.

You simply follow a high-fat, low-carbohydrate meal plan for five or six days a week. Then on day seven, you increase your carbohydrate intake to roughly 150 grams. On these days, focus on carbs like sweet potatoes, squash, and white rice. For reference, one sweet potato has roughly 115 grams of carbohydrates. If you are always in ketosis, your cells get "lazy" because they never burn glucose, and you actually develop insulin resistance.[12] To get the many benefits of ketosis without developing insulin resistance, sometimes eat a low-carb, ketogenic diet and other times eat a moderate carb diet.

After years of following a cyclical ketogenic diet, I now spend a lot less energy counting carbs. Instead, I use Brain Octane Oil because studies show it raises blood ketones even in the presence of carbohydrates. This way I can eat more veggies, enjoy a few carbs, and still have the benefits of ketosis. I put it in my Bulletproof Coffee, add it to my salad dressings, and drizzle it on meat so I get a small, steady dose of ketones throughout the day. My cells are always prepared to burn glucose or fat, and I avoid blood sugar spikes. As a result of eating this way, my insulin sensitivity score was a perfect 1 on a scale that goes up to 160 based on a combination of four laboratory values.

Cycling in and out of ketosis this way retrains the mitochondria in your brain to become resilient and metabolically flexible and gives your IDE a chance to stop breaking down insulin and work on cleaning house of cognitive-impairing plaques. Having ketones present in the bloodstream also reduces your levels of progranulin, the damaging protein released by your microglial cells when you are chronically inflamed.[13]

If you want to take your brain hacking up a big notch or you're dealing with early Alzheimer's, you can use insulin intranasally for cognitive enhancement. A compounding pharmacist pours a vial of injectable insulin into a nasal spray bottle, and you take one squirt in each nostril. I do this once or twice a month when I'm working on a project that requires an intense focus, like this book. It is useful as a cognitive enhancer in a healthy brain. There is also clear data supporting its use for Alzheimer's, as it improves delayed memory and keeps general cognition from declining.[14] It works for men and women, but the benefits top out for women at 20 IU, while men benefit from a larger dose of 40 IU. Check with your doctor about this biohack. Nasal insulin does not normally lower blood sugar—it just makes the brain better at using glucose as fuel.

Let me be clear: You do not want high levels of insulin in the brain for long periods of time because that will lead to insulin resistance, but quick bursts of insulin can be helpful to drive glucose metabolism. The supplemental insulin shunts glucose out of the bloodstream and directly into your neurons, giving them a hit of energy. The long-term consequences of having chronically high levels of insulin in the brain probably outweigh the benefits.[15] But I'd like to see more research on the effects of quick bouts of insulin. At this point it's safe to say that it's much more efficient to avoid dementia and Alzheimer's by keeping your insulin levels low through eating properly and exercising than it is to try to halt or reverse the disease later by adding exogenous insulin to your brain.

YOUR BRAIN ON DRUGS

In 1997 I stumbled upon a newsletter called *Smart Drug News* that had been around since the 1980s. Steven Fowkes, a biochemist, wrote and edited the newsletter. He hailed the benefits of a certain class of smart drugs called racetams, which includes piracetam, phenylpiracetam, and aniracetam. These drugs, which have been around since the 1960s, raise oxygen in the brain and improve mitochondrial dysfunction after oxidative stress.[16] The first positive studies came out in 1971, yet these pharmaceuticals were almost entirely absent from the United States. My doctor didn't know what they were, even though a major pharmaceutical company manufactures them. They weren't banned in the United States, but they weren't embraced, either, putting them in a "gray zone" where they still reside today.

Since I was desperate to get my brain back, I decided to take a risk and order $1,000 of these so-called smart drugs from Europe. When they arrived in a sketchy-looking unmarked package, I wondered if I was going to have a day as a genius or if I'd been conned out of a thousand bucks. Given the research, I considered it a good bet. I was right. Twenty years later, I still use these drugs every day.

Since that first dose, I have been experimenting with all kinds of smart drugs (also called nootropics) and supplements to boost my cognitive function. Some of them have had a huge impact on my brain function, and others have done very little. But I am grateful that I've had the opportunity to try them all and hope we reach the point soon where these drugs come out of the shadows and into the mainstream. We use coffee to perk up, glasses to see better, and all sorts of other tools every day to enhance our performance. There is no reason why smart drugs should be any different, especially the ones that make you smarter *now* and keep your brain working better for *longer*.

There are plenty of pharmaceuticals and supplements that can help you enhance cognitive function as you age. Some of the most promising anti-aging drugs trigger your body's production of naturally occurring chemicals (brain-derived neurotrophic factor, BDNF; nerve growth factor, NGF; and neurotrophin-3 and -4) that tell your brain to grow new neurons. Increasing levels of these chemicals can help treat degenerative brain diseases and boost your cognitive performance right now.[17]

It's hard to say which nootropics will provide you with the greatest return on investment, in part because everyone's brain is different and in part because we don't yet know everything about how they work. The most cautious approach is to simply try them one at a time. The bad news is that if you try each one for only sixty days, you'll be dead from old age before you try them all.

A far more effective strategy is to choose a result you're looking for and take several supplements at the same time that will likely provide that result. If you get it, you win. Then you can back off from some of the supplements to see if you still experience the same benefits. With pharmaceuticals, including the piracetam family, it's better to try them one at a time because pharmaceutical interactions are far more likely than with supplements.

Here's an overview of some of the smart drugs that can help keep your brain running well at any age. You don't need to run out and buy any of these, but it's worth your time (and money) to find one or two that provide the most benefit.

PIRACETAM

I first took the piracetam that arrived in that first unmarked package for two weeks and didn't feel much of a difference in my brain. I was mad. Those drugs were expensive, and I expected to see results. So I stopped taking the piracetam, and the next day in a meeting I found myself scrambling to think of a word. I suddenly realized that for the last two weeks, I hadn't fumbled like that. It felt so natural when the drugs were working that I didn't notice it. I had felt more like myself, and everything was just a little bit easier. There was no dramatic thunderbolt and I didn't gain superpowers. I just had a slightly greater capacity.

There are about a dozen other derivatives of piracetam, and each works differently. Try them individually, not as a "stack," until you know how your brain responds. One favorite is aniracetam, the only fat-soluble racetam, and the only one to increase memory I/O (in animals),[18] which is the ability to get things into and out of your memory. It is also a mild antidepressant.[19]

Another favorite is phenylpiracetam, which is banned in professional athletics because it increases physical performance. It is arguably the most stimulating of the racetams, and I wash it down with coffee when I want to really get something done. There isn't much evidence that phenylpiracetam makes young people smarter, but there is good evidence that it reduces cognitive decline in aging.[20]

I have taken aniracetam and phenylpiracetam on a regular basis for almost twenty years and plan to continue for at least the next hundred years. Normal doses are 500 to 750 mg of aniracetam twice per day, and 100 to 200 mg of phenylpiracetam two to three times a day. Ask your doctor about any possible drug interactions. Some people require extra choline, a B vitamin, with these compounds.

MODAFINIL

The next drug I tried was modafinil in 2002, and this time I felt the effects right away. It was like someone turned on the lights in my brain. It simply didn't take as much effort to use my brain as it did before. Modafinil helped me finish my MBA despite cognitive dysfunction while I simultaneously worked full time. It also improved my meditation practice. Without it, I don't believe I would have been able to start Bulletproof while working full time and being an effective husband and father of two young kids.

Very few entrepreneurs used smart drugs or even believed it was possible until a few years ago. The handful I knew would not admit to it publicly. But I was open about it from the beginning. I never wanted someone to talk about it later and accuse me of "cheating," so I made my use of smart drugs well known. At Wharton, I once lined up my nootropics on my desk before taking an exam. I even mentioned it in my LinkedIn profile, which led more than a few Silicon Valley friends to open up to me about the fact that they had used smart drugs, too.

Because I was the only guy willing to talk about it on air, ABC's *Nightline* came to my house to film a special on modafinil. Since then, modafinil has become more and more well known among entrepreneurs, executives, and even college students looking for an edge. The last part makes me sad—I don't think it's a good idea for healthy people under twenty-five to use modafinil because the prefrontal cortex isn't formed fully until that age, and there are no studies showing the effects it might have.

Today there's a good deal more evidence to back up the effects of modafinil, though there are reports that it increases alcohol sensitivity, so please don't drink while using it. Modafinil has been shown to increase your resilience and improve your mood. In healthy adults, it improves fatigue levels, motivation, reaction time, and vigilance. It even improves brain function in sleep-deprived doctors.[21] And it certainly helped me reverse my cognitive dysfunction.

I stayed on modafinil for almost ten years, until my brain was working so well that I felt I didn't need it anymore. In that time, I became a bit of a modafinil evangelist, constantly singing its praises to anyone who would listen. To write one of my very first blog posts, I sat down with some friends, including a successful television producer, a top artificial intelligence researcher, a published author, and a hypnotherapist. I shared some info about how much modafinil had helped me, and they all decided to get some with prescriptions or legal mail order. The next week, as I expected, I received some excited phone calls.

In one night, the TV producer finished a proposal for the Dalai Lama Foundation that he had been procrastinating on for months. He believed the proposal was far better than it would have otherwise been. The AI expert said he was able to make new connections he hadn't made before and suggested modafinil should be widely available. The author powered through his writer's block and made more

his current book than he had in months. And my hypno-
end felt she had huge breakthroughs in cognitive performance and made new connections in her mind on a technique she
was perfecting.

Honestly, these responses were not unusual. The only reason you
don't hear about them in mainstream outlets is that people are worried they will be seen as "cheating" or somehow weird. Well, I admit
it. Like our caveman friend who learned how to make fire and keep
his family warm, I cheated. And I'm weird, too. Those are common
Super Human traits!

MICRODOSE NICOTINE

More recently, I've been using oral nicotine for cognitive enhancement. Hear me out. I am not talking about cigarettes or vaping or any
tobacco product. Nicotine is only one of many chemicals in cigarettes,
and by itself it is a smart drug with few side effects.

A 1988 pilot study demonstrated the pronounced effects of nicotine on the brain in Alzheimer's patients.[22] After six patients received
intravenous nicotine, cognitive tests revealed decreased memory impairment, as well as fewer mood-related disturbances like anxiety and
depression. More recently, a double-blind pilot clinical trial revealed
that six months of a 15 milligram daily dose of nicotine is beneficial
for those with milder forms of cognitive impairment.[23] And the body
of research is growing. Nicotine may also help those with Parkinson's
and Alzheimer's due to its ability to work as an antioxidant in the
brain.[24]

Nicotine affects peroxisome proliferator-activated receptor gamma
coactivator 1-alpha (better known as PGC-1 alpha), the master regulator of mitochondrial biogenesis. This means that nicotine actually
helps grow new mitochondria. This is the same anti-aging mechanism as exercise! In fact, one reason so many people gain weight when
they quit smoking is because of the decrease in PGC-1 alpha. Yes,
you read that right: Nicotine creates some of the same changes in
cells as exercise. (Do both!)

I first became aware of this research in 2014 and have been using low-dose nicotine (1 milligram, which is about 5 to 10 percent of what's in a cigarette) for cognitive enhancement ever since. Along the way, I had to ask myself if I was okay with being addicted to nicotine—after all, it is an addictive substance—and I decided that I am. What if I told you I did something every day that made me feel good and if I stopped doing it I felt worse? You might call me an addict. The problem is that I'm talking about exercise! I don't consider it a weakness to be addicted to something that helps me perform better. I've come across lots of people who are only comfortable being addicted to the same things as everyone else, like air and water and maybe coffee. That's their prerogative, but it might mean they'll age just as quickly as everyone else, too.

Like modafinil, nicotine is not appropriate for anyone under twenty-five. You need a fully baked brain before you start messing with this stuff. And keep in mind that if you use nicotine, you'll have to check a box on your life insurance saying that you use it, or else abstain for ninety days. Also, it's important to take a minimal dose—while low doses can be helpful for aging and cognitive performance, high doses of nicotine are harmful for mitochondria and may even cause hair loss or erectile dysfunction.

While it hasn't been studied, the strategy that makes the most sense is low-dose (1 to 2 mg in divided doses) occasional use for cognitive enhancement starting at twenty-five and moving up to 1 mg two to three times per day up to age fifty, and adding another 1 mg per day approximately every five to ten years. In other words, when you're seventy, you might be up to 10 to 12 mg. That's about what you'd find in one cigarette.

If you do decide to try nicotine, be careful which products you use. Don't smoke or vape. Use oral (spray, gum, or lozenge) products instead. I'm a fan of the start-up Lucy, which makes nicotine products with clean ingredients and no industrial artificial sweeteners.

Let me be clear one more time: Smoking will not make you Super Human, help you live longer, or ever provide you with a positive ROI. Vaping is better than smoking, but the ROI is also negative. Just don't do it.

MICRODOSE DEPRENYL

One of the most powerful anti-aging smart drugs available is selegiline, also known as deprenyl. It is mostly known for stimulating production of dopamine, an important neurotransmitter that is involved in emotions, pleasure sensations, and the brain's reward and motivation mechanisms. Dopamine also helps control movement, which explains why nerve cell damage leading to a dopamine deficiency causes symptoms of Parkinson's disease such as tremors and loss of balance.[25]

There is a fine line between too much dopamine and too little. Too little is clearly problematic, but too much is associated with some pretty severe symptoms such as aggression and paranoia. Luckily, the amazing human brain has a mechanism in place to keep your dopamine levels in check. You naturally produce an enzyme called monoamine oxidase B, or MAO-B, that eats up extra dopamine in the brain.[26] If you don't have enough MAO-B, your dopamine levels climb, but if you have too much MAO-B, your dopamine tanks. This can leave you unmotivated, unable to feel pleasure, and withdrawn. On top of that, excess MAO-B puts surrounding cells in harm's way because its process of destroying neurotransmitters releases aging free radicals.

Unless you're dealing with neurological or psychological issues, the MAO-B checks-and-balances system normally serves you well until about age forty-five, which we typically think of as our peak. Around then, your MAO-B levels begin to rise year over year, which means dopamine starts to break down faster than you can replenish it. This is why most elderly people unfortunately have low dopamine levels.[27]

Selegiline/deprenyl works by blocking the enzymatic activity of MAO-B, which slows this breakdown of dopamine. To treat early-stage Parkinson's disease, doctors prescribe high-dose selegiline in pill form along with dopamine precursors. Together, they make extra dopamine and block the enzyme that destroys it.[28] There are also selegiline patches that doctors use to treat depression.[29]

In addition to blocking MAO-B, selegiline increases neurotrophic factors, compounds that strengthen existing neurons and support the growth of new neurons. Selegiline also increases superoxide dismutase, a powerful antioxidant that breaks down harmful substances in cells. This helps prevent tissue damage that can lead to hardening of the arteries, heart attack, stroke, and other inflammatory conditions.[30] These two effects are the reason I've been a fan of microdose deprenyl for twenty-two years.

Doctors have known about the longevity benefits of selegiline since the 1980s. Back then, a handful of animal studies showed that giving rats selegiline led to measurable life-span increases.[31] Rats that were given selegiline also became better learners,[32] and one study showed that giving rats selegiline restored behaviors that are typical of younger rats.[33] That's why I like it!

Keep in mind that selegiline interacts with other psychoactive medications and some over-the-counter medicines like dextromethorphan (an ingredient in some cough suppressants) and the plant medicine ayahuasca. At the normally prescribed doses, there are also possible physical and mental side effects such as nausea, sleep disturbances, impaired movement control, changes in heart rate, confusion, and more. These stem from having too much dopamine in your system. If you're starting with good dopamine levels, are young, and are supplementing with excess selegiline, you might experience side effects without benefits.

This is a prescription drug, and a functional medical doctor or anti-aging doctor will usually prescribe 1 mg per day (about a tenth of a high dose) starting in your thirties, and increase the dosage by 1 mg for every decade of age after that. In addition to the anti-aging effects, many users notice positive changes in motivation, energy, and concentration.

COENZYME Q10/IDEBENONE

Your mitochondria use coenzyme Q10 to produce energy. We all naturally have coenzyme Q10 in our bodies, but when you have extra free radicals and oxidative stress (you know, aging!), your mitochondria will use up your coenzyme Q10, leaving you with a deficiency. In addition, several pharmaceutical drugs, such as cholesterol-lowering statins, can reduce blood levels of coenzyme Q10 by up to 40 percent.[34]

Supplementing with coenzyme Q10 can help offset this decrease and provide you with more energy by helping your mitochondria work better.

If you plan to live longer than you're supposed to, CoQ10 must be on your short list of supplements at 100 to 200 mg per day. Advanced anti-aging experts often recommend idebenone, a synthetic pharmaceutical that is similar to coenzyme Q10 and is shown to improve skin and help brain cells stay healthy[35] and improve learning and memory in mice.[36]

PYRROLOQUINOLINE QUINONE (PQQ)

This antioxidant is about a hundred times as powerful as vitamin C at protecting your cells from free radicals to keep them young. It also stimulates NGF, which helps you grow new neurons, and has been found to enhance the regeneration of peripheral nerves that connect the brain and spinal cord to the rest of the body.[37] You can find PQQ naturally but not in useful levels in many foods such as green tea, fermented soy (natto), spinach, parsley, and (sadly, often inflammatory) green peppers.[38]

Research on mice supports PQQ's ability to kick mitochondria into high gear. Specifically, it can increase mitochondrial density to provide more energy,[39] reduce inflammation,[40] boost metabolism,[41] combat oxidative stress,[42] improve fertility,[43] improve learning and memory ability,[44] and protect the heart.[45]

PQQ also activates PCG-1 alpha in the same way that exercise and nicotine do, which sparks mitochondrial biogenesis.[46] This means that one supplement can enhance your existing mitochondria and help you grow new ones, all while acting as an incredibly powerful antioxidant. It's pretty much the holy grail of longevity. So why is no one talking about it?

There are two types of PQQ: stabilized disodium salts and active PQQ. Several years ago, I started taking 30 to 40 mg of the salt form of PQQ every day. However, unlike when I took other mitochondrial energizers, I never felt any effect. It's possible that it was doing something, but I couldn't detect a difference. PQQ is an expensive supplement, and I'm now convinced I was wasting my money and time on it.

The likely reason I didn't feel any energy even from large doses over an extended period of time is that it is sold as the "stabilized" disodium salt form because it's more convenient for the manufacturer. Unfortunately, in humans, disodium salts precipitate when they are exposed to even small amounts of stomach acid. That means that all that expensive PQQ I took was turning into little rocks in my stomach instead of helping my mitochondria. To get around that, I wrapped PQQ molecules in a protective coating of oil called a liposome to help them absorb. That is how the Bulletproof supplement ActivePQQ was born.

If you don't use the liposomal form, you could try taking PQQ salts on an empty stomach, possibly with some baking soda to neutralize stomach acid. There are no studies supporting this technique, but I bet it works.

L-THEANINE

This is an amino acid found in green tea that increases BDNF,[47] the growth factor that makes your brain more plastic. On its own, L-theanine promotes relaxation,[48] alertness, and arousal.[49] L-theanine also works synergistically with caffeine, so it's pretty convenient that it's found in green tea. Together, the two increase reaction time, memory, and mental endurance.[50] You can supplement with L-theanine or drink a cup or two of green tea per day.

If you decide to try green tea, look for one that's grown in the shade. Shade-grown green tea typically has much higher levels of chlorophyll, amino acids, and L-theanine than other varieties. This also increases the amount of caffeine in the tea and the sweetness of its flavor.

LION'S MANE MUSHROOM (*HERICIUM ERINACEUS*)

This staple of traditional Chinese medicine supports the brain and nervous system and promotes mental clarity, focus, and memory. These mushrooms are high in antioxidants and stimulate NGF. In fact, an isolated biopolymer in lion's mane was found to be more effective than NGF or BDNF in protecting neurons from oxidative stress.[51] I take this in combination with the other supplements that boost NGF and BDNF for the highest possible ROI.

There are different ways of extracting lion's mane mushrooms, and hot water is not a great one. That's why I don't recommend lion's mane capsules or tea, and it tastes horrible in coffee. The most effective form I've ever tried is a double extract that uses both alcohol and heat, made by Life Cykel. Two droppers before bed create noticeable increases in my REM sleep with powerful dreams I can easily remember.

CURCUMIN

A 2018 study[52] out of UCLA confirms that a daily dose of curcumin—the active ingredient in turmeric—improves memory and mood in people with age-related memory loss. In the double-blind, placebo-controlled study, forty adults between the ages of fifty and ninety who complained of memory issues were assigned to one of two groups. Group one received a placebo, while group two received 90 milligrams of curcumin twice daily for eighteen months. All forty participants took standardized cognitive assessments at the study's inception and then at six-month intervals. Thirty of the participants also underwent positron emission tomography (PET) scans to monitor brain amyloids before starting and then again after eighteen months.

The results revealed that the participants who took curcumin experienced markedly improved memory and attention abilities. In fact, the people taking curcumin improved their memory scores by an average of 28 percent over eighteen months. This group also noted mood improvements, and their brain PET scans showed less amyloid buildup.

To reap curcumin's maximum longevity benefits, take it in supplement form. Pair it with bromelain (a digestive enzyme found in pineapple), or take it in an oil-based capsule to increase your body's ability to absorb and utilize curcumin. Do not fall for the mistake of using black pepper or bioperine to increase absorption. Black pepper extract absolutely does raise your levels of turmeric and many other polyphenols. The only problem is that it does it by interfering with cytochrome P450 3A4 liver detox, which you need to stay young. That liver pathway cleans out pollutants, and by messing with that detox pathway, black pepper extract prevents your body from clearing potentially harmful compounds. So you end up with higher levels of turmeric and higher levels of aging toxins, too. This is not a good strategy. Black pepper extract has also been linked to leaky gut syndrome,[53] so I strongly recommend skipping it.

When I created the Bulletproof curcumin formula, I paired curcumin with a little-known Chinese herb called stephania root and frankincense to further help with inflammation, and I used oil-based capsules for absorption instead of black pepper. (Yes, I invest in creating things I believe in!)

HE SHOU WU (POLYGONUM MULTIFLORUM)

This ancient Chinese herb originally appeared in Taoist texts as a longevity enhancer. Now we know why. It stimulates the body to produce superoxide dismutase, an incredibly powerful antioxidant. It also inhibits MAO-B, increasing dopamine levels in the body[54]—kind of like deprenyl.

I started taking He Shou Wu a few years ago because of its ability to restore hair growth and reduce grays. Yes, most of my anti-aging hacks focus on feeling rather than looking young, but it's okay to want to look good. Most of my relatives went gray before they were thirty, and I noticed my own hair starting to turn gray when I was in my thirties, along with a slightly receding hairline. There are a handful of studies to show that He Shou Wu helps rats regrow hair,[55] and it's been considered one of the most effective ways of restoring color to gray hair since ancient times.

In fact, He Shou Wu translates to "He's black hair." According to legend, the man who discovered this herb took it every day and went from being completely gray to having a full head of lustrous black hair. He also allegedly lived in good health to a hundred and sixty. Since taking He Shou Wu, I have noticed my grays reducing, but I also take other supplements that help. I expect to regrow a full head of black hair *and* maintain a fully functioning brain when I'm a hundred and eighty.

Bottom Line

Want to age backward? Do these things right now:

• Keep your blood sugar stable, even after meals, by reducing sugar intake. Extra sugar causes oxidative stress you don't want. If you do eat sugar or carbs, pair them with fiber or saturated fat. Extra points for adding chromium and vanadium supplements to control blood sugar spikes.

• Develop metabolic flexibility with a cyclical ketosis diet to fuel your brain and keep your neurons from becoming insulin resistant.

• Make sure you have low levels of ketones present in your body much of the time, but not all the time. I use Brain Octane Oil, especially when I eat carbs.

• Consider neurofeedback treatment if you've experienced significant trauma or suffer from symptoms of anxiety or depression. You've got to get that fixed, or it will tax you for the rest of your life.

• Try a cognitive enhancer from the list in this chapter to promote healthy brain function and avoid cognitive degeneration as you age. Here is the short list:

  • Piracetam: Reduces cognitive decline with age
  • Modafinil: Performance enhancing, not anti-aging
  • Nicotine: Low doses (not from cigarettes) can be helpful for aging and cognitive performance
  • Deprenyl: Works on dopamine receptors for cognitive enhancement
  • CoQ10: Helps your mitochondria produce energy
  • PQQ: A powerful antioxidant for anti-aging
  • L-theanine: An amino acid that helps with memory and mental endurance
  • Curcumin: Improves memory and attention while acting as an antioxidant
  • He Shou Wu: Longevity-enhancing antioxidant herb that can also help you regrow and regain color in your hair!

METAL BASHING

When I was in my mid-twenties and suffering from symptoms of aging ranging from arthritis to cognitive dysfunction, I was lucky to find a doctor who was experienced in identifying heavy metal toxicity. I tested positive for high levels of mercury and lead and probably had significant amounts of other metals in my body, too. This may sound shocking, but really the only surprising thing was that I knew about these metals, not that I had them. If you're alive and reading this there is a very high likelihood that you have dangerous levels of toxic metals in your body, too. Even worse, "safe levels" of toxic metals aren't safe at all if your goal is to get younger.

If you're young and resilient, you may not yet feel the effects of heavy metals. But make no mistake; if they are present in your body, they are impacting you in subtle ways. Imagine again that you're Superman, and someone has sprinkled kryptonite everywhere on Earth. It's in the soil that grows your food, in the water supply, and so on. Little by little, you consume small amounts of this poison. It doesn't kill you right away—it just lessens your powers. Every time you ingest it, you're a little bit weaker.

This is quite literally happening with heavy metals. We consume this poison little by little, and it weakens us—causing invisible cellular damage while suppressing our immune system and thyroid function. As a result, we feel slightly more sluggish and foggy with each passing year. We say that it's just age catching up with us, but it's actually the slow and steady creep of metals building up in our bodies and taxing our systems, aging us unnecessarily.

THE POISONOUS EFFECTS OF METALS

Arsenic, cadmium, lead, and mercury are the most toxic and present metals in our environment. Although the EPA has classified each of them as carcinogens,[1] today we are consuming them in considerable quantities. More mercury is present in our food now than ever before, and other heavy metals such as aluminum, nickel, thallium, and even uranium often appear in high concentrations in our bodies. In addition, copper, iron, chromium, and zinc are essential nutrients in the body, but at high levels they, too, are toxic and prevent your cells from functioning optimally.

Heavy metals are part of the Earth's crust. Human activities such as mining, smelting, and manufacturing—and in some countries, the continued use of lead in paint, gasoline, and aviation fuel—have brought them out of the crust and into our soil, air, and drinking water. Even worse, sewage sludge from cities is contaminated with high amounts of heavy metals—enough to be considered low-level toxic waste. Private companies controlling waste disposal commonly mix the sludge into fertilizer until it's diluted just enough to meet EPA limits. Then they spread the metal sludge onto the fields used to grow your food, and the food takes it up. Then your cells do.

You already know that our mitochondria produce energy via an electrical process. Well, when these metals, which have high electrical conductivity, get in our bodies, they mess with that process, causing a dramatic increase in oxidative stress. This has a direct impact on cellular function,[2] leading to premature aging and decline.

Of course, we know that babies and children are particularly vulnerable to the toxic effects of metals. In children under the age of ten, metals can cross the blood-brain barrier and kill off neurons, leaving these children with cognitive and mental health problems as well as reduced IQs. According to the World Health Organization, when pregnant women are exposed to high levels of metals, the unfortunate results are miscarriage, stillbirth, premature birth, and low birth weight, which can lead to lifelong health problems.[3]

We are still learning about all the ways that metals contribute to aging in adults. In 2018, *The Lancet* released a study that examined the connection between lead exposure and deaths from cardiovascular disease. They analyzed data from over fourteen thousand adults, correcting for variables like age, sex, ethnic origin, location, smoking, diabetes, alcohol intake, and even household income. The results were shocking, revealing that adults with the highest levels of lead exposure had a 70 percent increase in mortality risk from cardiovascular disease and *double* the mortality risk from coronary heart disease.[4] This means that with all else being equal, high levels of lead exposure increases your risk of death from this killer by 70 to 100 percent.

This is because when lead enters the blood vessels, it damages the cells that line them, hardening arteries and causing plaques to form. Once plaques are present, blood pressure increases, as does the risk of heart disease and stroke. While government agencies have consistently reduced acceptable amounts of lead in the environment since the 1970s, *The Lancet* concluded that "there is no safe threshold for lead exposure." And yet, only twenty years ago it was deemed perfectly safe to have levels of lead present in the blood that we now know to be toxic. It pays to be more conservative than government safety standards, which are always influenced by economics and don't put your health and longevity first.

Another growing but often overlooked toxic metal is thallium, which is used in rat poison and other insecticides, and which the electronics, glass manufacturing, and pharmaceutical industries use in production. Thallium is called "the poisoner's poison" because Russian spies have used it to assassinate people. It is colorless and tasteless, and it replaces potassium in your cells, making them fail. Unfortunately, in a shockingly shortsighted move, the oil industry decided to replace the lead in gasoline with thallium, which is way more toxic. Even in smaller doses, it causes degenerative changes in many organs. These adverse effects are the most severe in the nervous system,[5] as thallium poisoning can cause lesions in part of the basal ganglia. Damage to this part of the brain leads to problems with speech, movement, and posture.

Sadly, because of its ubiquitous presence in U.S. soil and fuel, thallium is hiding in plain sight in one of the world's trendiest vegetables: kale. As kale consumption and harvesting has exploded over the last decade, so has thallium exposure. We've known for years that kale and other brassica vegetables such as cabbage are exceptionally good at taking up thallium from soil. A 2006 peer-reviewed paper by Czech researchers confirms this to be true of kale,[6] and a 2013 study from China found the same issue in green cabbage.[7]

In fact, brassicas are so effective at soaking up thallium that in 2015 Chinese researchers found they could use green cabbage to purify soil of thallium.[8] In other words, the cabbage soaked up all the thallium in the soil, leaving the soil itself toxin-free. Think about that the next time someone offers you a kale smoothie or coleslaw made with conventionally grown cabbage!

Mercury is another common toxin that builds up in your tissues over time. Scientists have clearly established that it causes high blood pressure, cardiovascular disease, and neurotoxicity.[9] In other words, it makes you more likely to die from two of the Four Killers: heart disease and Alzheimer's. Mercury can also cause you to experience impaired cognitive function and motor skills. This one is particularly troublesome because it's found in fish from the ocean, and we must eat fish to get adequate omega-3 fats. But if you eat a lot of fish, you're getting a lot of mercury.

Fish are an awesome metaphor for our own bodies. The older the fish, the more heavy metals have bioaccumulated in its tissues. That's why eating a large halibut, swordfish, or shark is a bad idea. These fish can live for more than a hundred years, and every time they eat a small fish that contains mercury, the mercury sticks around in the tissue of the larger fish. Likewise, if your plan is to live to beyond a hundred with your faculties, you're not going to want a hundred years of accumulated mercury, lead, and other metals hanging out in your brain.

But one of the largest sources of mercury exposure in humans is one we've created ourselves by using mercury amalgam in "silver fillings" for dentistry. They're so full of mercury that when a silver filling comes out of your mouth it must be treated as hazardous waste. If you have lots of mercury fillings, there is a 100 percent chance you will not perform at your potential or live as long as possible. It's that clear. The problem is that removing these fillings improperly will cause a huge spike in your blood mercury levels, likely poisoning your brain in the process. That's why a mercury-free dentist who knows how to remove fillings safely is a must-have on your Super Human support team.

Compact fluorescent light bulbs are another common source of mercury. Each bulb has enough mercury vapor to create a hazardous waste cleanup problem according to government safety limits. I've taught my kids to literally hold their breath and run out of a room should a bulb break at school or a friend's house. We don't allow them in our home.

The bottom line is, no matter how clean you think your food or your environment is, you've undoubtedly been exposed to enough heavy metals to accelerate your aging and increase your risk of succumbing to the Four Killers.[10] And the longer you live, the worse it gets. So, the question is: What are you going to do about it? Will you plead ignorance and do nothing just because your doctor never tested you for metals, or will you take action to proactively avoid and reverse the aging these metals cause?

HEAVY METAL DETOX METHODS

Exposing yourself to fewer of these toxins is a good strategy, one you must follow in order to live longer and become Super Human. The only problem is that many of them are omnipresent, and you simply can't avoid some degree of exposure. That's why it is essential to periodically detox from heavy metals and purge them from your system.

The practice of detoxing has some negative connotations, mostly due to a mix of addiction treatment "detox" centers, random laxative teas, and other "cleanses" full of sugary fruit juice and maple syrup that are somehow supposed to help you purge toxins. As a result, many people turn up their noses at the very concept of detoxing. I get it. But there is a big difference between someone wanting to make a buck by capitalizing on the idea of detoxing and our actual need to help our bodies remove harmful and aging substances.

Other people resist the idea of detoxing because they believe that our bodies naturally eliminate anything that's potentially harmful. Perhaps this would be possible if we all still lived in the Garden of Eden and we got our food (and toxins) only from Mother Nature. But that ship has sailed, and a high toxic load from our modern environment makes it tough for your body to clear toxins efficiently. You simply weren't engineered to live a long time while swimming in the modern soup of man-made chemicals and toxic metals more present in our food supply and in our bodies than Mother Nature ever intended.

If your digestive system is healthy and you don't overload your system with an even higher than average toxin load, you will most likely eliminate the majority of the metals you ingest through your stool. But a small percentage of heavy metals are stored in your fat cells. If you're familiar with compound interest, you know that a small percentage of interest on your savings account compounded for fifty years can create real wealth. Likewise, a small percentage of heavy metal accumulation compounded for fifty years or more will create real aging and biological chaos.

I'm actually grateful my symptoms of toxicity were so bad that I pursued heavy metal testing because it encouraged me to learn about it and start clearing these aging substances from my body. If you have symptoms of heavy metal toxicity or your health is crappy in general, see a functional medicine practitioner who can evaluate you and guide you through a safe detox process. Detoxing metals the wrong way can be dangerous—it's possible to accidentally move metals from your tissues into your brain. A functional medicine practitioner will order either a urine test or a hair test for heavy metals. Hair tests are easy to do. They don't require you to pee in a bottle or take drugs that release the heavy metals you're storing in your tissues, so they're harder to interpret and don't measure your total body burden of toxic metals. Urine tests provide a more accurate picture of your toxin levels and are considered the gold standard.[11]

Whether you have proof of your own heavy metal exposure or not, I encourage you to take action to slowly eliminate metals from your body as you age in one or more of the following ways.

GLUTATHIONE AND OTHER ANTIOXIDANTS

Back in 1999, the first thing I did to detox from heavy metals was to get an IV of glutathione. This was far less popular and more expensive than it is now. Glutathione is one of the body's most powerful antioxidants, and it can protect you from heavy metal damage.[12] Glutathione is also a natural chelating agent, meaning its molecules can form several bonds to a single metal ion. As you read earlier, chelating agents bind to the metals in your body, inactivating them. Then your body eliminates the metals through urine and bile. Glutathione is particularly helpful because it keeps mercury from entering cells in the first place.[13]

Unrelated to heavy metals but worth noting: Glutathione protects fats from oxidation, supports your mitochondria, boosts immunity, and helps your brain function at its peak.[14] Glutathione also recharges other antioxidants, making them more effective at fighting inflammation, and is a cofactor for dozens of enzymes that neutralize damaging free radicals.[15] The lower your levels of glutathione are, the higher your risk of all the Four Killers climbs.

Of course, your body makes its own glutathione, but it's hard for any of us to produce enough of it to keep up with increased free radical production as we age. Add heavy metal exposure to the mix, and there's a clear case for most of us to take additional glutathione. The first time I got an IV of glutathione I felt massively better right away. From then on, any time I felt like I was getting sick, I went to my doctor and got an IV of glutathione, and it really helped. To this day, I continue to make sure my glutathione levels remain high by taking supplements and occasionally getting glutathione via IV after a long flight. I skip glutathione doses sometimes so my body doesn't naturally downregulate its own production. Taking it every day for long periods without a break isn't a good idea.

Another powerful antioxidant that helps chelate your heavy metals is alpha-lipoic acid (ALA). This antioxidant can cross the blood-brain barrier to protect the membranes of neurons from heavy metal damage.[16] ALA also regenerates old glutathione both inside and outside of cells, increasing glutathione levels in the body.[17] In addition, ALA helps mitochondrial performance. Some researchers believe that you must take ALA every four hours to avoid redepositing metals in your brain, but most physicians I know do not follow that protocol. We do know that oral doses of up to 1,800 mg per day have no side effects.[18]

It's always a good idea to supplement your detox efforts with vitamin C, the world's best-known antioxidant, since low vitamin C levels are associated with low levels of glutathione and excess oxidative stress.[19] Like ALA, vitamin C recycles used glutathione, elevating antioxidant levels in red blood cells.[20] And by itself, vitamin C can help you detox from lead.[21] As with glutathione, I skip vitamin C on some days and for twelve hours following high-intensity workouts because the oxidative stress created by a workout is part of the signal to your muscles to grow. If you take it after workouts, vitamin C will disrupt that signal.[22] In addition, adequate zinc levels prevent your body from absorbing lead and cadmium,[23] which is why I take a mixed zinc and copper orotate capsule daily as a part of my anti-aging program.

ACTIVATED CHARCOAL

Another low-hanging fruit of detoxification is activated charcoal, a form of carbon that has a massive surface area and a strong negative charge. Activated charcoal has been used for more than ten thousand years by Chinese medicine healers, Ayurvedic practitioners, and Western medicine doctors alike. It's still used in emergency rooms today to treat poisoning.

Charcoal works through a process called *adsorption*, which means to bind to rather than to absorb. In the body, it binds to chemicals whose molecules have a positive charge. Once the charcoal attaches to these chemicals, you can pass them normally (i.e., poop them out). Many toxins, even those made naturally by bacteria and toxic mold, will bind to charcoal, so you can excrete them before they harm your body.

When you eat foods containing cadmium, copper, nickel, and lead (but unfortunately not mercury), activated charcoal can bind up these toxic metals before they have a chance to stick to your cells.

Charcoal can help prevent many of the cellular changes associated with aging. In one study, activated charcoal increased the life-span of older test animals by an average of 34 percent.[24] Even if the increase is less dramatic in humans, this is a relatively risk-free intervention with clear anti-aging effects. In fact, a 34 percent increase in life-span is unprecedented from any pharmaceutical. Remember, an increase in life-span is an increase in the maximum amount of time you can live. It's far more difficult to extend life-span than it is to increase the average amount of time you can expect to live.

Scientists have known about the effects of activated charcoal on heart health since the 1980s. In one study, patients with high cholesterol who took activated charcoal three times a day showed a 25 percent reduction in total cholesterol and doubled their HDL/LDL cholesterol ratios.[25] Yet almost no conventional doctors recommend activated charcoal to their heart patients.

When I first saw this research many years ago, my only experience with charcoal had been on the slopes of Annapurna in Nepal, where it was sold everywhere because it relieves symptoms of almost any gastrointestinal problem. When I returned to the United States, there were very few capsules of activated charcoal available on the market. So I bought charcoal in powdered form, mixed it into a beaker full of water, and grimaced as I chugged down the bland-tasting, gritty drink. But the next morning, I woke up noticeably less puffy and more focused. Then I went to the bathroom and thought I was dying because the charcoal passes through your GI tract and turns your poop black, the same color as blood in your stool. Consider yourself warned.

Keep in mind that the activated charcoal in supplements you'll find on the market can come from a variety of sources. Charcoal is the result of burning something, and in the case of some charcoal supplements, that thing is farm waste. There are also many different grades of charcoal, ranging from the coarse stuff found in water filters to ultra-fine particles. I prefer (and manufacture) activated charcoal made from the husks of coconuts washed with acid to dissolve any heavy metals present in the coconut husk and then ground into the finest possible particles. The finer the particle, the greater the surface area to bind toxins once ingested. One gram of activated charcoal has a surface area ranging from 950 square meters up to 2,000 square meters.[26] In other words, fine-grade charcoal is twice as effective as normal grade. In fact, finer particles are proved to bind to the most carcinogenic substance known to man, the mold toxin aflatoxin.[27] I take these capsules almost every day on an empty stomach as part of my overall anti-aging strategy and as a way to continuously detox from chemicals, pesticides, and some heavy metals.

If you go this route, remember to never take activated charcoal at the same time as prescription medications or some other supplements. Charcoal binds a lot of substances—even the good stuff like prescription medications, vitamins, and minerals. Wait an hour or more after taking charcoal to take other supplements or medications. Talk to your doctor about the details. If you're on antidepressants, this is even more important. You don't want to "detox" your antidepressant before it takes action in your brain!

CHLORELLA

In animal studies, a type of algae called chlorella binds very well to mercury in the gut,[28] and a great many physicians recommend consuming it for this reason. I have noticed a measured difference in my fine motor control and neurological function when I eat chlorella along with fish, a common source of mercury. A best practice is to take twenty-five or more chlorella tablets with meals containing fish. This way, you can get the precious omega-3 DHA from fish without taking a hit from the mercury.

MODIFIED CITRUS PECTIN AND OTHER FIBERS

Mother Nature's original method of detoxing harmful substances from your gut is to use simple digestive fiber. There are two kinds of fiber: the insoluble kind that you can't digest and the soluble kind that feeds good bacteria in your gut. It's important to feed them because simply having healthy gut bacteria can help you detox.[29] You'll read more about this later.

In addition to feeding your gut bacteria, it turns out there is a form of citrus fiber called modified citrus pectin (MCP) that has almost magical anti-aging powers. It's good at removing lead, cadmium, arsenic, and thallium. In one study, about 15 grams of modified citrus pectin powder per day for five days caused the study subjects to pass significantly higher levels of metals through their urine. Specifically, the amount of arsenic leaving the body increased by 130 percent, cadmium levels increased by 150 percent, and lead levels increased by 560 percent.[30]

That alone is good for aging, but there's more. MCP reduces cancer's ability to spread in the body.[31] Given that cancer is one of the Four Killers, taking a substance that makes it harder for cancer to kill you so you have more time to kill it only makes sense.

MCP also reduces levels of a molecule in your body called galectin-3. This helps protect you from harmful bacteria that cause local inflammation and immune activation. As you age, your levels of galectin-3 go up, which creates chronic inflammation.[32] You've already learned that chronic inflammation leads to an increased risk of all Four Killers and is tied to the formation of AGEs, one of the Seven Pillars of Aging. Galectin-3 is also associated with heart failure, kidney disease, and cancer.[33]

However, galectin-3 is essential for young people (under forty) to grow healthy tissues, so unless high levels of metals are present, they shouldn't take modified citrus pectin every day. I give my kids, who are nine and twelve, about 5 grams once a week. A low dose for adults is 5 grams per day for several months, and a high dose is 15 grams per day for a year. My pragmatic strategy is to take 15 grams every other day for a year, and then 5 grams every couple days as a maintenance dose. It has a mild taste, so when I take it, I add it to my Bulletproof Coffee or just stir it into a glass of water in the morning. Given the benefits for continuous metal detox and the reduction of heart disease, cancer, and kidney risk, this one is worth your time.

EDTA CHELATION THERAPY

The big gun of chelation therapy is ethylenediaminetetraacetic acid (EDTA) chelation, which was first used in 1950 to treat lead poisoning. EDTA is a synthetic amino acid that binds to metals and diminishes their reactivity. This helps you remove metals and reduce the damage they can cause, and it is used to reduce calcification of your arteries as you age. EDTA is also an anticoagulant for blood. Can you picture the small amount of a yellowish substance at the bottom of the vial that keeps your blood from clotting when you have it drawn? That's usually EDTA.

After my first glutathione IV in 1999, I then took the next step to detox from heavy metals: EDTA chelation therapy, also via IV.

I moved to the big guns quickly because lab tests showed my blood was so sticky that I was at a high risk of heart attack or stroke. I wasn't yet thirty years old. When I got home from receiving the EDTA chelation, my significant other at the time looked at me and said, "Wow, your skin is so pink!" I had never even noticed that for years my skin had been an unhealthy gray color. The EDTA therapy helped my blood flow better, finally bringing some color to my cheeks. Since then, I've also tried suppositories of EDTA chelation and found that they worked just as well, if not better, for continued detoxification. You may not like the idea of suppositories, but when you compare going to a doctor's office and sitting for an hour to get EDTA via IV with the fifteen seconds it takes to insert a suppository, it's pretty obvious which one requires less of an investment of time and money.

EDTA is a potent chelating agent for calcium as well as heavy metals. As we age, we typically build up calcium in our tissues, and this calcification leads to all sorts of symptoms of aging, from heart disease to even balding (more on this later). This calcium buildup is usually the result of an imbalance between calcium and other vitamins such as vitamin D and vitamin K_2. Your body needs vitamin D to properly absorb calcium, and almost everyone today is deficient because of our strict avoidance of UV light.

We need vitamin K_2, on the other hand, to keep calcium in its place—namely, our bones and teeth. Too often, calcium leaches out of the bones and teeth as we age, leaving behind little holes, and then the calcium builds up where it doesn't belong: in soft tissues. When this happens, EDTA chelation is extremely effective at pulling the calcium out of your tissues. If your body isn't yet calcified, supplementing with vitamins D and K_2 should help you avoid tissue calcification and protect your teeth and bones from aging. Lead is also tied to calcification, so EDTA can be doubly powerful when it removes calcium and lead![34]

When my father had calcification issues a few years ago, I recommended EDTA chelation, and it worked incredibly well. Mainstream cardiologists have been claiming for years that EDTA chelation doesn't work, but that doesn't change the fact that many functional medicine doctors successfully use EDTA chelation to improve arterial function. If you are over forty, I recommend getting a calcium score to measure your calcification levels. If your levels are elevated, EDTA chelation either via IV or suppository can assist in lowering them while helping you detox from heavy metals, which you undoubtedly have, too. If you are under forty or unable to get a calcium score, you can try one of the milder chelation therapies above preventatively.

As with all other medical chelating agents, I recommend getting tested for heavy metal buildup before using EDTA chelation. Even if you don't choose to do a full urinary metal lab test, talk to a functional medicine doctor before trying EDTA chelation. You really want to go to a medical professional for this one, as you can become seriously ill if your liver and kidneys aren't able to process the metals once they've been released from fat cells. When done correctly, however, chelation therapy can rapidly decelerate and even reverse aging. It's worth the investment of time and money to work with a professional.

SWEAT IT OUT

Your body does have its own way of detoxing that doesn't require any IVs or supplements: sweat. Sweating does more than cool you off. It also helps you get rid of heavy metals and xenobiotics, foreign compounds such as plastics and petrochemicals, in small but significant amounts. A 2012 systematic review of fifty studies found that sweating helps remove lead, cadmium, arsenic, and mercury, especially in people with high heavy metal toxicity.[35]

I recommend stacking your benefits by getting your sweat on in an infrared sauna. Whenever I feel that I've been exposed to something toxic, such as when I eat in a restaurant and can tell the food was high in contaminants, I do a one-hour infrared sauna session to detox. Keep in mind that sweating pulls electrolytes and trace minerals from your body, so it's important to drink a lot of fluids and get plenty of salt (preferably Himalayan pink salt or another mineral-rich natural salt) if you're going to use a sauna to detox.

Of course, you can also get a sweat going the old-fashioned way: by exercising, which increases lipolysis (the breakdown of fat tissue). This releases heavy metals stored in your fat tissue. I recommend doing high-intensity interval training (HIIT) once or twice a week to boost lipolysis. Exercise has plenty of other anti-aging benefits as well. Research shows that adults who regularly engage in intense exercise have significantly longer telomeres, those protective caps on the ends of chromosomes that you read about earlier. As a result, people who exercise regularly are a full decade younger than their peers on a cellular level.[36]

Remember that mobilizing toxins is a good thing only if your body is able to actually get rid of them. You want to be sure you're expelling toxins, not just moving them to a different part of your body! Working out improves circulation, providing more oxygen to your liver and kidneys so they can better filter out toxins once they've been released from fat cells. But lipolysis is even more effective when you combine it with supplements that support the liver and kidneys. This includes calcium-D-glucarate, which converts to glucaric acid in your body and supports the crucial detoxification pathway in your liver. Like activated charcoal, glucaric acid scavenges your body for toxins and then binds to and eliminates them so they can't age you prematurely. It's also a good idea to take activated charcoal before you work out or hit the sauna to suck up the toxins released from fat cells.

Another way to benefit from lipolysis, this time without breaking a sweat, is by getting into a state of nutritional ketosis. Ketosis is a very effective way to induce lipolysis, particularly when fasting. When you're in a fasted state of ketosis, your body breaks down your fat stores so it can produce ketones for fuel. Since heavy metals are stored in fat cells, this means you can supercharge your detox (and fat loss) by dropping into nutritional ketosis. As you've read, that comes from either fasting for two days or eating mostly fat, moderate protein, and almost no carbs. If you try this, boost your detoxification efforts by mopping up with activated charcoal, calcium-D-glucarate, or both.

I occasionally do longer fasts to reap the detoxification and anti-aging benefits. It may sound difficult at first, but once you become fat adapted (meaning your body is accustomed to breaking down fat for fuel), it's really pretty painless. Plus, I love saving time by not having to do dishes for a few days. The ability to fast without feeling like you're going to die is a Super Human power that you can grow.

Bottom Line

Want to age backward? Do these things right now:

- Up your antioxidant levels with supplements that will help you detox and counter the negative effects of metals in the body. Focus on glutathione, alpha-lipoic acid, zinc orotate, and good old vitamin C.
- Regularly bind the metals you are exposed to by taking activated charcoal, 500 mg to 5 grams per day, and/or modified citrus pectin, 5 to 15 mg per day, both away from food or pharmaceuticals. Take some chlorella tablets when you eat fish.
- If you feel you are aging faster than you'd like or have a reason to believe you've been exposed to high levels of heavy metals, see a functional medicine doctor to get your urine levels tested. If they are indeed high, consider IV chelation therapy or suppository EDTA chelation therapy under a doctor's supervision.

POLLUTING YOUR BODY WITH OZONE

In 2004, I went to an unlikely place to reverse all the things that had gone wrong with my biology: a dentist's office. Through my anti-aging nonprofit work, I met an eighty-eight-year-old dentist named Dr. Gallagher who was based in Sunnyvale, California. Dr. Gallagher had found that many people unknowingly had low-grade infections in their teeth, setting off horribly aging widespread inflammation. And when dentists placed caps on top of those infections, they festered, slowly causing aging inflammation.

In his practice, Dr. Gallagher used ozone therapy to sterilize teeth before placing caps on them. Ozone therapy isn't just for teeth, however. Dr. Gallagher had learned about ozone therapy from Frank Shallenberger, who pioneered the field of ozone therapy more than forty years ago. But ozone therapy has been around even longer than that. It evolved as a treatment for infection before we had antibiotics. German doctors first used it to successfully treat wound infections during World War I.

Why is it, then, that you probably don't know much about it? When most people hear the word *ozone*, they think of the ozone layer, a region of the Earth's stratosphere that absorbs the majority of the sun's UV rays and contains high concentrations of ozone. In the 1970s, scientists became concerned that pollution from chemicals in the atmosphere was depleting the ozone layer, leading to dangerous increases in UV radiation. They responded by banning certain chemicals, and the ozone layer is recovering. Ozone is also the name for an air pollutant that causes health problems on smoggy days. Between these two associations, the idea of "ozone therapy" sounds . . . crazy.

That didn't stop me from visiting Dr. Gallagher—not for help with my teeth or an active infection, but to see how he might help reverse my many symptoms of aging, especially those stemming from my obesity and toxic mold exposure. Dr. Gallagher took me under his wing. This kind, wise old man was full of energy and more than willing to share his enthusiasm for a forgotten technology that works against almost every infectious agent and even restores energy to broken mitochondria. He set a prime example of what the wisdom of old age has to offer to the world. And after one treatment of ozone in my ear canals, which gets oxygen into the brain, I felt so good that I was sold on the benefits of ozone therapy as an anti-aging technology. I've been using it regularly ever since and have been honored to invite the man behind modern ozone therapy, Dr. Frank Shallenberger, onto *Bulletproof Radio* and onstage at my annual biohacking conference.

OXYGEN AND OZONE: THE BIG OS

Right now you're breathing. Well, at least I hope you are. By simply inhaling, your body is taking in oxygen, which in its molecular form contains two oxygen atoms (hence the name O_2). Ozone, on the other hand, has three oxygen atoms. You can think of it as O_3. Because ozone has an extra molecule, the atom itself is very unstable, or reactive.

Based on everything you've read so far, this sounds like a bad thing, right? Oxidative stress caused by excess reactive oxygen species (free radicals) is a major cause of aging. And indeed, ozone can be damaging when used incorrectly. Ozone is a colorless, very strong-smelling gas. If you breathe it in low quantities, you'll cough and wheeze. At higher quantities, you'll experience uncontrolled vomiting, and above that, you can die from lung damage.

So why the heck would anyone in his or her right mind pursue ozone therapy? Well, ozone can be safely administered intravenously, rectally, vaginally, or transdermally (through the skin). You can also drink ozonated water. This is a type of hormetic stress. It sends a signal to your body that weak and dysfunctional cells should die and the rest of them should get stronger. Weak cells that are vulnerable to invasion from bacteria or viruses are more susceptible to oxidation. Ozone therapy kills off these weak and damaged cells.

At the same time, ozone destroys harmful bacteria, yeast, viruses, fungi, and protozoa. Ozone does this by oxidizing lipids in the body, creating lipoperoxides. These peroxide bodies inactivate pathogenic bacteria, viruses, fungi, yeast, and protozoa in different ways. With fungi, the peroxide bodies inhibit growth. With viruses, they disrupt virus-to-cell contact, preventing the virus from reproducing. And to kill bacteria, these peroxide bodies oxidize the bacterial cell wall, destroying the bacteria almost immediately. Ozone therapy kills 99 percent of bacteria within seconds and is a hundred times more effective at killing bacteria than bleach! Ozone therapy is also more effective than antibiotics and is highly preferable because it kills bacteria without wiping out your beneficial gut bacteria or taxing your immune system.

In fact, ozone therapy *strengthens* the immune system. The signal that peroxide bodies are on the loose triggers your body to naturally produce more of its two most powerful antioxidants, whose names should be familiar by now: glutathione and superoxide dismutase. Ozone therapy also leads to a significant increase in the production of interferon, a protein that inhibits virus reproduction, and two signaling molecules of the immune system: tumor necrosis factor and interleukin-2.[1] An increase in these signaling molecules leads to a cascade of subsequent immunological reactions, powerfully boosting the entire immune system.

Research shows that ozone therapy can literally be lifesaving. When researchers preconditioned rats with ozone and then injected them with lethal fecal material, the rats' survival rate went from *zero* to 62.5 percent.[2] Ozone is also effective in treating antibiotic-resistant bacteria,[3] something I expect to see a lot more of in the future thanks to our widespread overuse of antibiotics.

When I first met Dr. Gallagher, he took me to the back of his practice and within half an hour taught me how to use ozone safely at home. Then he said, "Here's the medical gear I have. It costs fifteen hundred dollars." I was crestfallen. That was more money than I could reasonably afford to spend at the time. "But this," he continued, "is the Chinese gear that costs less than two hundred. The only problem is it's not metered, so you don't know how much ozone you're getting." He offered to have a friend of his mark it up for me, and I was all set.

I just had to find medical-grade oxygen, which was harder than it sounds. It turns out there's an incredible medical mafia around this stuff. So I did what any good biohacker would do, which is to go on Craigslist and buy a used welding oxygen tank, which holds exactly the same thing as medical-grade oxygen. I ordered a medical regulator that allowed me to control the flow rate of oxygen coming out, and I began my at-home ozone treatments for less than $500.

At home, I had two choices for treatment: rectal and vaginal. I'm only technically equipped for one of the two, so it really wasn't much of a choice at all. I put a small amount of ozone into a special little bag with a hose, squeezed it into my rectum, and went about my day. Being a professional biohacker isn't always glamorous.

The first time I did this, I could feel my brain turn on almost immediately. It was amazing. The burst of energy was so precious to me because I hadn't felt it in years, despite my best efforts. But it lasted for only five minutes. I knew I was onto something, so for the next eighteen months whenever I wasn't traveling, I did a little at-home rectal ozone treatment every night. This helped reverse the damage from years of toxic mold exposure and got me back to the healthiest state I'd been in for many years. It was so transformative that as soon as I could afford it, I invested in medical-grade ozone equipment.

You might be wondering why ozone therapy is in a book about living longer than you're supposed to. Here's why: It is one of the most effective ways to restore your mitochondrial function to that of a young person and to stop any infection that could kill you on your way to immortality. It's also very affordable and shockingly safe compared to almost any pharmaceutical.

Since learning how to use it, I've turned to ozone therapy whenever I have a local infection that I want to nip in the bud before it spreads and I have to take antibiotics. I even use it on my kids. A few years ago, my then eight-year-old daughter, Anna, scratched her ear on a rosebush, and it got seriously infected. Her ear was hugely swollen and an angry red color. Dr. Lana (a former ER doctor) decided that we would take Anna to an urgent care clinic for antibiotics if it didn't clear up by the next morning. Knowing how bad antibiotics are for your gut (more on this later), we did what any good biohacker parents would do. We sat her down next to the medical ozone generator, put on a TV show about dragons (a special treat, since she doesn't watch a lot of TV), blew a fan on her so she wouldn't breathe in the ozone, and funneled the ozone gas directly onto the skin on her ear. After the first treatment, her ear returned to its normal size. After the second, it returned to its normal color and was completely healed.

Why don't more doctors know about this treatment? It takes some time to administer ozone therapy, and it requires training, but it doesn't sell drugs. And because it's not the dreaded "standard of care" forced on healers by insurance companies, any doctor that uses it is taking on some amount of risk.

Thankfully, there are some doctors who are willing to take that risk. For example, Dr. Robert Rowen. I first met Dr. Rowen when he presented at the Silicon Valley Health Institute and later interviewed him for *Bulletproof Radio* about the work he did training doctors in Sierra Leone to treat patients suffering from Ebola with ozone therapy. In most cases, patients with Ebola have only a 10 percent chance of survival, but Dr. Rowen knew he could improve those odds if he trained the local doctors to administer ozone therapy. Not only is ozone safer than other treatment options for Ebola, but it's also less expensive and far more effective.

In October 2014, Dr. Rowen traveled to Sierra Leone with his colleague Howard Robins after being invited by the president to train local health professionals to treat Ebola with ozone therapy. The training went great, but then they hit a snag. While they were at the Sierra Leone Ebola treatment center outside of the capital, a call came in from the Ministry of Health halting the ozone project with no further explanation. Patients and staff were forbidden from receiving ozone therapy, even if they knew they'd been exposed to Ebola. The staff did continue with the training, fearing for their lives. But patients were denied the treatment and left to die.

Four health care workers and three doctors who were on the front lines subsequently contracted the disease. Of the three doctors, two outright refused ozone therapy and died. This made international news. The third doctor had already been trained by Dr. Rowen and requested ozone therapy but was refused. Sadly, he also died. The four health care workers, however, managed to get the ozone therapy, and all four responded nearly instantly. They completely recovered within a few days with no complications. Dr. Rowen published the results of those four cases in the *African Journal of Infectious Diseases.*[4]

The wife of one of the senior doctors had encouraged her husband to get ozone treatments, and after his death, she was placed under armed guard quarantine at her home. She knew she had been exposed to the disease and was afraid for her life, but the guards wouldn't let her leave to get ozone treatment. She became so desperate that she scaled a razor wire fence in the middle of the night, shredding her skin, in order to escape. Luckily, she was able to receive the ozone therapy and developed no symptoms.

I am tremendously grateful for brave and disruptive doctors like Dr. Rowen, who take on these risks to bring effective treatments to the people who need them. It's shameful that greed has kept more people from benefiting.

OZONE THERAPIES

As you read earlier, there are a variety of ways to get ozone into our bodies. The fastest and most effective option is to expose your blood to ozone. This way, the ozone goes straight into your bloodstream and instantaneously reacts with your blood atoms. To do this, the doctor takes blood out of the body, places it into a container, and then injects the ozone in its gas form directly into that container. The ozone gas quickly disappears when it comes into contact with other atoms, creating peroxides. The container that once held blood now holds peroxides and white blood cells. When infusing these cells back into your body, you expose your entire system to the benefits of ozone.

Today you can spend even more to go to a doctor's office and get the best systemic ozone treatment around, which is called a ten pass. For this treatment, a doctor pulls out your blood, mixes it with ozone, and puts it back into your bloodstream. Then he or she pulls out your blood again, mixes it with ozone, and puts it back in, ten times over and over. This powerful treatment is a substantial procedure, but it does wonders to rejuvenate the body and remove microbial pathogens. I plan to do ten-pass ozone once a year for at least the next hundred years. It might be a good fit for you if you're suffering from a long-standing illness or just want to reset your mitochondria to become more youthful.

You can also use ozone locally either transdermally or via injection, which causes the peroxides to quickly interact with your cells in an infected area. For example, if you have chronic cystitis (inflammation) of the bladder, you can inject ozone directly into the bladder. When you use ozone rectally, the peroxides drain straight to the liver. This can be used as a direct liver treatment. It's also possible to "ozonate" water and then apply the water to specific areas to treat local infections. Some dentists recommend swishing with this water to fight off gum disease, and you can use the water as eye drops to treat eye infections. Please don't try any of these therapies before consulting a doctor! This bears repeating: Accidentally inhaling ozone gas can cause permanent lung damage or even kill you. Don't go it alone without basic training.

More recently, ozone has become a popular option for treating chronic joint pain, an all-too-common symptom of aging. A promising therapy called prolozone involves injecting ozone directly into the joint, where it triggers healing and regeneration. I learned about this from a member of SVHI who was in her fifties, weighed about four hundred pounds, and was diabetic. One month, she relied on a walker to get around. The next month, after receiving prolozone therapy, she needed only a cane. Medical images of her knees showed no visible cartilage before the treatment and a clear layer of cartilage just six weeks after receiving prolozone therapy. At the time, scientists believed you could not grow cartilage! If chronic joint problems, especially in your knees and back, are a part of your life, this might be a game changer for you. I have since worked with doctors to inject ozone into two of my joints with considerable improvements.

Today I opt to rotate my oxidative therapies because I find that each one has its own unique benefits. Sometimes I do IV UVB, as I mentioned earlier, and on occasion I use a high-dose vitamin C drip to change things up. Ozone is my go-to when I want to keep my energy levels and immunity high after a long flight, and via IV as an occasional anti-aging treatment.

Think of it this way: Excessive oxidation causes your body to basically rust from the inside out. This is why excess free radicals are bad for you. It's counterintuitive to imagine that adding an extra oxidant like ozone to your body will help, but it forces your body to become better at making its own defenses. It's like lifting weights for your cellular antioxidant systems!

OZONE, ENERGY, AND NAD

I now know that my first at-home ozone treatment turned my brain on, even temporarily, because ozone is also one of the world's most powerful mitochondrial stimulants. It increases the rate at which red blood cells break down glucose to create energy while boosting the amount of oxygen being sent to tissues. But perhaps most important, it charges a powerful anti-aging molecule called nicotinamide adenine dinucleotide (NAD) with its extra electron, transforming it into NAD+.

NAD is a coenzyme (a compound that certain enzymes need in order to work) that every one of your cells needs to perform its basic functions and keep you alive. Unfortunately, your levels of NAD decline by 50 percent between birth and age fifty,[5] and they continue declining even further after that unless you do something about it. To understand what NAD does, it's helpful to think of NAD as a waiter that picks up an electron from one table and drops it off at another. It exists in two forms: NAD+ and NADH. The oxidized form, NAD+, grabs an electron from one molecule. While it has a hold on that electron, it becomes NADH. NADH then donates that electron to another molecule, and it becomes NAD+ again. This simple act of shuffling electrons around allows your enzymes to activate microscopic chemical reactions in your cells that keep them healthy and your whole body humming. And ozone therapy adds extra electrons to the party.

Without NAD+, everything grinds to a halt. Your mitochondria require NAD+ to create energy. Your body also needs NAD+ for the maintenance of blood glucose levels at night, for the generation of ketones, for proper muscle function, and for your nerves to send messages to one another. In a nutshell, you cannot function without it. In one study, when scientists created conditions in the lab that inhibited NAD+, cells died because their mitochondria couldn't make enough energy to sustain life.[6]

NAD+ also helps ensure that proteins retain their shape so you can avoid the buildup of amyloid proteins, one of the Seven Pillars of Aging. And your cells use NAD+ to help a particular family of proteins called sirtuins function. Sirtuins regulate biological pathways and protect your cells from oxidative stress and its resulting age-related decline.[7] These proteins also play a crucial role in maintaining the length of your telomeres.[8] Because you want long telomeres and the longevity that goes along with them, you need plenty of NAD+.

There's plenty of research to show that NAD+ itself also protects cells from oxidative stress. In one study, when researchers measured a cell's NAD+ levels, they were able to predict whether or not it would survive in the face of a stressor. The more NAD+ a cell has, the better its chance of survival.[9] In another study out of Singapore, researchers stressed rat brain cells by depriving them of oxygen and then added NAD+ directly to the cell culture. They found that the cells treated with NAD+ were far more resilient to stress. A much higher percentage of the cells that received the NAD+ survived than those that did not get the extra shot of NAD+.[10] Translation: Healthy levels of NAD+ keep you and your cells strong and resilient, especially in the face of stress.

NAD+ also helps your body repair DNA that has been damaged from normal day-to-day exposures and functions.[11] It does this by bringing a negative charge to places where DNA is damaged, which facilitates repair.[12] This has been shown to extend the life-span of animals. In one study on rats, a control group that did not receive NAD+ supplements all died within five months. Half of them died even sooner, within three and a half months. Meanwhile, the group that received supplemental NAD+ all lived beyond ten months, indicating that NAD+ supplementation led to a dramatic improvement in life-span.[13] If this molecule can help a rat live 50 percent longer than expected, what can it do for you?

Your ratio between NAD+ to NADH plays a big role in how you'll age. Ideally, you want a NAD+ to NADH ratio of 700:1. Most healthy people have this proper ratio in their cells until around the age of forty, when NAD+ levels begin to decline.[14] It's no big surprise that this coincides with increased oxidative stress and rapid cellular aging.[15] Athletes are usually the first ones to notice an NAD+ decline when they suddenly can't bring it the way they used to despite keeping up with their usual training regimen. Other "normal" people usually don't notice that something has changed until another decade or so has passed. Then they chalk it up to simply getting older, and sadly have no clue that they can enhance their NAD+ to NADH ratio and begin aging backward instead of quickly sliding downhill.

There are actually lots of things you can do to keep your NAD levels from declining and even raise them to youthful levels. This is a good idea regardless of your age. Raising NAD+ levels increases insulin sensitivity,[16] reverses mitochondrial dysfunction,[17] reduces stem cell senescence, and even increases life-span in animals.[18] How's that for taking on all Four Killers and shoring up two pillars?

If you want to pull out all the stops, you can opt for intravenous NAD+ treatments. This was originally used as a treatment option for drug and alcohol addiction, but it's now a regular part of my quest to live to a hundred and eighty. It's also part of the human upgrade program at Upgrade Labs in Beverly Hills. To date, I have completed twenty full IV treatments, although the standard anti-aging regimen is five and the drug and alcohol regimen is ten. Each one takes about ninety minutes, and the first few are intense. It may sound strange, but in a way it reminded me of the first time I tried ayahuasca in Peru, but without the hallucinations. I could feel a strange, tingly pressure in my chest, which bordered on uncomfortable, and then I felt warm all over. I slept incredibly well after my first NAD+ treatment and felt great for days afterward. From now until the time I reach a hundred and eighty, I plan to continue getting IV NAD+ treatments once a quarter to supercharge my mitochondria and keep myself from ever peaking.

Here are some other ways to boost NAD+ levels:

- Take supplements that will boost your NAD+ levels. I like Tru Niagen, which contains the only known FDA-safety notified form of nicotinamide riboside, a precursor to NAD+.

- Follow a cyclical ketogenic diet. Whether they come from fasting, restricting carbs, or using Brain Octane Oil to raise ketones directly, ketones increase the NAD+ to NADH ratio.

- Practice intermittent fasting and/or calorie restriction, which both increase NAD+ levels.[19]

- Take oxaloacetate (a part of Bulletproof's KetoPrime supplement). Your body converts this organic compound to malate, a salt form of malic acid, which raises your NAD+ to NADH ratio.[20]

Bottom Line

Want to age backward? Do these things right now:

• If you're over fifty or have dealt with serious health problems in the past, find a local ozone doctor and get IV treatments when they are affordable for you. At worst, your mitochondria will become better. At best, the ozone will knock out other unpleasant stuff growing in your body that you don't even know about.

• If you have arthritis or sore joints that don't get better, consider prolozone injections into the impacted joint to speed healing dramatically.

• If you're having dental work done, look for a dentist who uses ozone gas to sterilize the teeth before treatments. This can help you avoid chronic inflammation and its corresponding aging.

• Up your NAD+ with supplements or IV treatments to boost mitochondrial function at any age. If you don't want to try either of these, you can increase your NAD+ levels through cyclical ketosis, intermittent fasting, and/or calorie restriction.

9

FERTILITY = LONGEVITY

As you read earlier, I was twenty-six when I had my first hormone tests done at an anti-aging physician's office. When the results came back, I learned I had low testosterone, way more estrogen than was normal (enough to grow man boobs), and exceptionally low thyroid hormone levels. In response, I went on prescription bioidentical testosterone and thyroid, and this changed my life. I can share firsthand how important these hormones are for how you look, how you feel, how you perform in bed (or even if you care about what goes on in bed), and how much you like your job and your life in general.

No matter how old you are, knowing your hormone levels and making sure they are ideal can only benefit you. If you want "normal" hormone levels for a seventy-five-year-old when you're seventy-five, go for it. I'll be rocking my twenty-five-year-old hormones when I'm seventy-five, either because by then I'll have upgraded my biology enough to produce hormones like a young man or because I'll still be supplementing with hormones to make me feel Super Human.

SEX, LIES, AND HORMONES

In 2001, as I set out to learn about the complex world of hormone replacement, I met T. S. Wiley at the American Academy of Anti-Aging Medicine. I was already familiar with her work. Her first book, *Lights Out*, informed a lot of my early thoughts on the importance of circadian rhythms, and her second one, *Sex, Lies, and Menopause*, completely changed the way I thought about hormones and fertility.

That convention was filled with the brightest stars of the anti-aging world, but most people outside those walls had never heard of these brilliant and badass doctors, authors, and researchers. At the time, those in mainstream medicine considered the entire field of anti-aging to be crazy. That's still true to some extent today. But within the field, Wiley is considered a maverick because she helps her older clients use hormones to mirror the natural hormonal cycles of much younger people. If normal hormone replacement is paint by numbers, her protocols are van Goghs.

When I saw Wiley's booth at the convention, I got in line right away to meet her. Directly in front of me was a gynecologist. When it was her turn to speak to Wiley, she raved about how good she felt on the Wiley Protocol, which is Wiley's recommended dosing of bioidentical hormones to keep postmenopausal women young.

Wiley smiled, looking younger than her years. "Oh, that's great," she said. "Do you use it with your patients?"

The doctor paused. "Oh, no," she said. "I can't do that. It's too complex, and the regulators wouldn't understand."

This was the first time I saw firsthand a doctor who knew about and benefited from a treatment she wouldn't share with her patients—and all because of federal regulations. Wiley was clearly frustrated, and so was I. To this day, it baffles me that it can be considered ethical to withhold useful and even life-changing information from your patients simply because it's not the standard of care.

That's exactly why I find it so ironic that people criticize Wiley for being a researcher instead of a medical doctor. Her recommendations for bioidentical hormone replacement therapy come largely from a detailed understanding of how hormones work combined with self-experimentation. She clearly knows her stuff. Her books are incredibly well researched and backed up by citations from medical literature, and the fact that she is not beholden to a regulatory medical association means she can actually try things and see what works without fear of losing her license. And take it from me, one can learn a lot about the human body from self-experimentation.

For some reason, though, the idea of hormone replacement is still somewhat controversial. People consider it unethical or dangerous to take hormones after your body has stopped producing them in ideal amounts. We buy into the myth that since our bodies naturally begin producing lower levels of some hormones as we reach middle age, the only option is to accept it and stoically face a swift decline. When our hormone levels go down, we don't just become infertile—our bodies begin to age in numerous, devastating ways.

Hormones are chemicals that serve as messengers in the body, relaying important information to various glands and organs. Specifically, hormones govern your thyroid, adrenal, and pituitary glands, as well as your ovaries or testicles and pancreas. In fact, hormones control much more than most people realize. When our hormones are out of balance, we feel it—and the list of symptoms is nearly endless.

But unless you have a serious health condition, if you have your hormone levels tested by a conventional doctor, he or she will probably tell you that you are within the normal range for your age. But the "for your age" part should be a huge red flag for anyone who isn't looking forward to the typical course of aging in our culture. I want my hormones to be within the normal range for someone who is the age I want to *feel*, not the age on the calendar. To be within the normal range of someone who is beyond his or her reproductive years means you are dying, albeit slowly. It may not be particularly comfortable to hear, but it's true. You can accept your fate and start sliding downhill at around the age of forty, or you can work to keep your hormones at the levels of someone at his or her peak so you can keep kicking ass long term.

In many ways, the natural reduction in hormone levels as you get older drives the aging process. In both sexes, as you age your body produces less and less pregnenolone and dehydroepiandrosterone (DHEA), two pre-hormones that your body converts into the other major sex hormones: testosterone, estrogen, and progesterone. With less DHEA, you also have less of these other hormones, and as a result you can expect more skin aging,[1] more body fat, less muscle, lower bone density, crappy sleep, and often, sexual dysfunction.

Today, after menopause women live about a third of their lives deficient in sex hormones. Even before that, two-thirds of women experience reduced sexual pleasure during the period of time directly preceding menopause, which is called perimenopause. Most women enter perimenopause in their early forties.[2] It's not just about sexual pleasure, though. Estrogen, which starts to dip during perimenopause, helps prevent cancer, bone loss, and heart disease. When estrogen levels decline, your risk of these killers rises. Men need estrogen, too, and they face the same risks when their levels decline in middle age.

Testosterone also does a lot more than you may think in the bodies of both men and women. This hormone helps both men and women burn fat, build muscle,[3] and maintain a healthy sex drive. Low testosterone is linked to osteoporosis in men, mild cognitive impairment, and even Alzheimer's. Yikes! You can consider bioidentical testosterone replacement a potent nootropic, or smart drug, because it improves verbal, spatial, and working memory and executive function.[4]

Testosterone is also lower in men who have heart attacks, suggesting that proper levels can help you avoid this major killer.[5] A study of eighty-three thousand older men who underwent testosterone replacement therapy found it decreased their risk of mortality from all causes.[6] Yet doctors routinely underdiagnose and undertreat the decrease in testosterone that men experience at around the same age that women enter menopause, called male andropause.[7] Remember the movie *Grumpy Old Men*? That's andropause. You can avoid it.

For men, Wiley recommends cycling between testosterone, DHEA, and growth hormone in a natural rhythm to stay young and avoid this fate, and to start as young as twenty-nine, since your levels will drop off about ten years later. This will keep your immune system strong and your inflammation levels low, adding quality years to your life. For women, she recommends using bioidentical hormones to mirror the exact hormonal cycle of a woman in her peak reproductive years. She has female patients who still menstruate in their sixties, and Wiley says that these women have made the choice to keep menstruating and stop aging. That is powerful.

END THE STIGMA OF HORMONE REPLACEMENT THERAPY

Yet, hormone replacement has a bad reputation for no good reason. In 2002, the Women's Health Initiative found that women who were treated with nonhuman estrogen from horse urine and a modified pharmaceutical form of progesterone had a higher risk of breast cancer, stroke, and cardiovascular disease.[8] That was enough for people to associate "hormone replacement" with cancer. But the same study also found a reduction in hip and back fractures and no total increase in cancer. In other words, the increase in breast cancer was offset by a decrease in other types of cancers.

Still, after the study came out, hormone replacement lost its luster. Within three months of the study, prescriptions dropped by about half. That may have been a good thing, however. It's not a great idea to use synthetic hormones, especially since bioidentical hormones that are exactly the same as what your body makes are available. However, bioidentical hormones cannot be patented, so they don't receive major research funding.

Meanwhile, testosterone replacement has a similarly bad reputation because of 1970s body builders who used synthetic testosterone pharmaceuticals to get jacked. Unfortunately, their livers also got jacked. To this day, when many people hear about testosterone replacement for men, they think of body building instead of staying young, lean, and vibrant at any age. Using bioidentical testosterone instead of synthetics makes a huge difference, as does dosing it for aging instead of body building.

There is also a myth that testosterone replacement will cause prostate issues, but this connection is based on old animal studies. Three recent studies failed to find any link between prostate issues and testosterone therapy, but two did find that low testosterone levels are associated with an *increased* risk of prostate cancer.[9]

So let's just get rid of preconceived notions about hormone replacement and think about what would happen if as you aged you maintained the hormone levels you had at twenty-five or thirty. Before trying any of this, though, it's best to get a lab test to learn your current hormone levels. You can order these tests online to save money, but if you need help interpreting the results or getting a prescription for hormone replacement, it's best to work with a functional medicine practitioner. Before prescribing any hormone replacement, a physician will measure:

- Estradiol—the most potent estrogen

- Estrone or estriol—weaker estrogens (estrone turns into estriol in your body)

- Progesterone—balances estrogen

- Testosterone—necessary for muscles and sexual function

- Prostate-specific antigen (PSA)—for men who are on testosterone, discontinue if it climbs

- Hematocrit—for men who are on testosterone, discontinue if it climbs

- Sex hormone binding globulin (SHBG)—this sticks to hormones to make them unavailable

- DHEA—a pre-hormone

- Pregnenolone—a pre-hormone

I've supplemented with many of these, including:

TESTOSTERONE

As you read earlier, I started hormone replacement therapy (HRT) when I was in my twenties and my hormones, particularly testosterone, were trashed. At first my physician prescribed a cream that I rubbed into my skin, which my body absorbed transdermally. This is the form of treatment that Wiley recommends, too. Personally, I found the cream to be inconvenient. I worried that if someone touched me, the testosterone would rub off on him or her. This can be dangerous with kids or pregnant women. If you're sleeping next to someone, the cream can get on the sheets and transfer over that way, too.

There is a dirty little secret about testosterone cream that almost no one knows, however, and I'm going to share it here. Please don't abuse it. If you are a woman and you take a vanishingly small dose of testosterone cream and apply it to your labia and vulva (or ask your partner for help), you will experience a form of vasodilation (increased blood flow) rarely seen no matter how good your partner is in bed. It has a profound local effect and will produce a night you won't soon forget. You don't want to do this every night, but on an occasional basis it is unlikely to be harmful and very likely to blow your mind. Compounding pharmacists call it "scream cream" for a reason.

And in fact, having better sex is also a great anti-aging strategy. There's real data to back that up. A ten-year study of nine hundred men found that men who had sex only once a month died twice as often as those who had it twice a week.[10] Given that this test started in 1979, as was the practice at the time, researchers didn't bother to ask about the women.

In the last few years, a lot of men and women have switched from a hormone cream to a pellet that goes under your skin. This is probably the best way to take testosterone now. The pellet is life-changing for both men and women. The regular dose for women is much lower than it is for men. Women, you won't get bulky and grow a beard when you take testosterone to achieve ideal hormone levels, but you will probably lean out a little without losing your curves, and your energy and sex drive will be amazing. Female body builders who experience weird, scary side effects are taking anabolic steroids, not bioidentical testosterone. For the pellets, you visit a doctor's office once every three to four months to get a new one.

I started bioidentical testosterone at age twenty-six and have been on it for most of the last twenty years, except in 2013, when I was testing the Bulletproof Diet to see what it did to my testosterone. (It raised my testosterone, but not as much as I wanted.) In that time, my sex drive returned, my energy increased, my brain function improved, I lost weight, and I gained muscle. Yes, like a young person—which my body thought I was. Please don't tell it the truth.

DHEA

The pre-hormone dehydroepiandrosterone (DHEA), which you read about earlier, has been getting a lot of attention lately for its longevity-enhancing effects. A major review of the available research on DHEA concluded: "During the past five decades, a myriad of animal experiments has suggested that DHEA is a multifunctional hormone with immune-enhancing, anti-diabetic, anti-obesity, anti-cancer, neurotropic, memory-enhancing, and anti-aging effects."[11] *Boom.*

Sadly, DHEA levels plummet as you age, reaching a mere 10 to 20 percent of youthful levels by the time you hit seventy.[12] These levels are even lower in postmenopausal women. When the ovaries stop producing hormones during menopause, the only source of sex hormones in a woman's body is her adrenal glands, which produce hormones in much lower levels than ovaries. Low DHEA is linked to depression and a higher risk of heart attacks and death from all causes. Low DHEA is also related to sexual dysfunction.[13]

You may read this and decide to just buy some DHEA supplements, but that is a bad idea. Without proper testing, you don't know what your body is going to do with those hormones once it gets them. I first tried taking DHEA when I was only nineteen. My man boobs got even bigger, and my libido tanked. It turns out that the men in my family have a gene that preferentially turns androgenic hormones, including DHEA, into estrogen!

As a side note, this genetic fluke causes a few other hormone therapies that work for some people to completely fail for me. For instance, men can take human chorionic gonadotropin (hCG), a pregnancy hormone in women, to stimulate the body to produce more testosterone. Some guys swear that it also causes penile growth,[14] but this was certainly not my experience. When I tried injecting hCG, I grew a nice round bubble butt, developed soft thighs, and regained the man boobs I had fought so hard to lose. My body literally thought I was pregnant. Every body is different, so it's incredibly important to find a doctor you trust, get the proper lab tests, and take any hormone supplements under supervision.

OXYTOCIN

Even oxytocin, a hormone best known for its role in making you feel good and bond with others, is easy to supplement with and has powerful anti-aging effects, specifically when it comes to tissue and muscle maintenance and regeneration. A 2015 study showed that inhibiting oxytocin reduced muscle regeneration, while supplementing with oxytocin caused rapid muscle regeneration via enhanced muscle stem cell activation.[15] In other words, the additional oxytocin told the body that it was young and needed to build more muscle, so the body quickly recruited latent stem cells for the job. Oxytocin also works alongside your inflammatory stress hormones such as cortisol during times of stress so you maintain homeostasis and don't become too inflamed.[16] Since reducing inflammation is pretty much the most effective anti-aging strategy of all, you clearly want to make sure you have plenty of oxytocin in your body.

Meanwhile, we know that our levels of circulating oxytocin typically decline with age. This is clearly one reason our stem cells become less activated as we get older and we become more inflamed. It also makes me wonder if it's why having strong relationships and being a part of a close-knit community helps people live longer. Perhaps in addition to offering psychological benefits, those relationships cause oxytocin levels to rise, and that oxytocin helps these people stay young. Another great benefit of oxytocin is that it reduces your stress response and dramatically boosts feelings of happiness. And you already know that less stress equals less aging.

As you read earlier, for most of my life I was more than a little bit socially awkward. So I was curious and hopeful when I learned about the benefits of oxytocin from Dr. Paul Zak, a neuroeconomist whose research on oxytocin has earned him the nickname Dr. Love. In his lab, Dr. Zak has conducted hundreds of experiments measuring oxytocin levels in participants' tissues and blood and manipulating it using oxytocin injections or nasal spray. He says that an increase in oxytocin makes people more generous and trusting, less suspicious, and better able to read social cues.

Learning this, I saw that it might be possible to hack my empathy and my aging at once. I had my oxytocin levels tested, and they were indeed far too low. The healthy range for men is 7 to 18 pg/ml, and I measured at a measly 4. So I got to work boosting my oxytocin levels using both natural and pharmaceutical means.

Any type of human interaction will increase oxytocin, but some are more effective than others. Face-to-face communication stimulates the greatest release of oxytocin. So I make sure to regularly schedule time with my team at Bulletproof in person, even though most of the staff works remotely. On a personal level, getting a massage, sharing a bed with a partner, playing with a dog, and even giving someone an eight-second (or longer) hug all boost oxytocin levels. I make sure to do all of these things on a regular basis.

I've also experimented with oxytocin lozenges and nasal sprays, which are available from a compounding pharmacy with a prescription. There are some retailers selling oxytocin products online, but be careful if you choose to go this route. Some are simply labeled "oxytocin" and are filled with other junky ingredients. I highly recommend working with your doctor if you want to try supplementation.

When I used the nose spray, I felt a bit happier and more at ease, but the change was very subtle—not dissimilar to how I feel after playing with my kids or my dog. Most of the time, I opt for the latter because it improves my performance while keeping me young and has the benefit of raising my kids' (and dog's) oxytocin levels, too. Of course, playing with my dog and kids has plenty of other mental, emotional, and even spiritual benefits for all of us. We do these things naturally because they feel good. But it's nice to know that they can simultaneously help you age backward!

HGH

Finally, human growth hormone (HGH), the famous anti-aging compound, declines as you age. Unfortunately, it is frighteningly expensive and doesn't have clear science backing its use as a supplement. If lab tests show you are deficient in growth hormone, you will most likely benefit from replacing it. If you want to just *look* young, you may still benefit, but if you want to actually live longer, the results are mixed.

I was clinically diagnosed with deficient growth hormone about two years before I started writing this book. My levels were exceptionally low, which raised my risk of the Four Killers. I take a low dose of HGH on occasion, but given that it is expensive and hard to travel with, I do not take it every day. Unless your lab tests indicate that you really need it, this one may not be the best use of your money.

However, if you are recovering from a major surgery, there's a case for using HGH to recover.

WHAT'S HURTING OUR HORMONES?

The process of aging itself causes a shift in our hormones, but it's not the only factor in the hormonal fluctuations that occur with age. Two common culprits of hormone imbalances are a poor diet and exposure to environmental pollutants. In other words, modern living is not kind to our hormones. In fact, American men have seen their average serum testosterone levels decline by about 1 percent each year[17] over the past several years.

Here's how your body makes testosterone: cholesterol→ pregnenolone→ androstenedione→ testosterone.

Testosterone begins with cholesterol. In fact, you synthesize every single sex hormone from cholesterol. This is one reason that a "heart healthy" low-fat, low-cholesterol diet is horribly aging. Research confirms that men who eat saturated fat, monounsaturated fat, and cholesterol have higher testosterone levels than those who follow a low-fat diet.[18] (Maybe there really is such a thing as benign vegan-induced testicular atrophy!)

Carbohydrates, on the other hand, deplete your hormones, specifically testosterone. You may find this hard to believe, but some common high-carb foods like cornflakes and graham crackers were invented a century ago to lower the male libido. Kellogg and Graham believed that male sexual desire was at the root of all society's problems, so they set out to make bland foods that would reduce libido. (This is true; look it up.) That low-fat grain-based thing absolutely works wonders for lowering testosterone, if that's your goal. It's certainly not mine.

There are two keys to naturally boosting testosterone through your diet: getting enough fat and getting the right kinds of fat. A study from back in 1984 looked at thirty healthy men who switched from eating 40 percent fat (much of it saturated) to 25 percent fat (much of it unsaturated), with more protein and carbs to make up the difference in calories. After six weeks, their average serum testosterone, free testosterone, and 4-androstenedione (an important hormone for testosterone synthesis) had all dropped significantly.[19] By the way, the idea that a low-fat diet was healthy began to catch steam in the mid-1970s, shortly before the nationwide testosterone decline started. It could be a coincidence, but I doubt it.

The other problem with a Western diet is that it's lacking key micronutrients that we need to create hormones, specifically vitamin D, which is essential for testosterone production. As you read earlier, almost everyone is now deficient in vitamin D because of our overavoidance of UV light. This is likely a major reason behind the decrease in testosterone levels. A study published in 2010 looked at the vitamin D and testosterone levels of more than two thousand men over the course of a full year. The results showed that men with healthy vitamin D levels had more testosterone and lower levels of sex hormone binding globulin (SHBG) than the men who were vitamin D deficient.[20] SHBG binds to hormones so your cells can't use them. If you have too much of it, your testosterone levels will drop.

The other interesting thing about this study is that the men's testosterone levels were lowest in March (at the end of winter) and highest in August (at the end of summer). Sunlight exposure affects your vitamin D production, and you are meant to have seasonal dips and peaks. Many of our hormones rise and fall in natural cycles. This is one reason hormone replacement therapy is such a tricky business. If you take a pill every day, you won't experience the natural rhythm of hormone increases and decreases that our bodies are meant to cycle through as we move through time. This is precisely why the Wiley Protocol for both men and women includes carefully created dosing cycles to replicate nature.

I recommend getting a blood test to check your levels of vitamin D, and if you're deficient, take a high quality vitamin D_3 supplement. If you're going to take D_3, take vitamin K_2 and vitamin A along with it, since they work together synergistically. While you're at it, take a look at your zinc levels, as a zinc deficiency can also cause low testosterone. If you're low on zinc, try eating more grass-fed red meat and/or taking a zinc orotate supplement.

In addition to the impact of the standard Western diet on our hormones, hormone-disrupting chemicals are now more prevalent in our environment than ever before. Many popular deodorants, lotions, shampoos, conditioners, shaving creams, and other grooming products now contain hormone disrupters, chemicals that either mimic the effects of hormones in the body or interfere with the function of your hormones. The worst offenders are phthalates, which, in addition to being a nightmare to pronounce, mimic estrogen and accumulate in your fat cells,[21] and parabens. All four common types of parabens—methylparaben, ethylparaben, propylparaben, and butylparaben—are estrogenic, meaning they bind to estrogen receptors and change the way estrogen in your body functions.[22]

For women specifically, there is another major cause of hormone disruption that has become taboo to talk about, but it would be remiss not to include it here: hormonal birth control. It is a basic human right to be able to use any technology or compound you want to control your biology, and that includes birth control in all its forms. However, some forms are better for your long-term health than others. The sad truth is that birth control pills are aging. They contain synthetic estrogen and progesterone. This diminishes pituitary hormone levels, which regulate many processes throughout the body. The result is not just the suppression of ovarian function, which stops women from getting pregnant, but also a decrease in testosterone production.

Women also need testosterone for sexual desire, to be sensitive to sexual touch, and to reach orgasm. In 2010, German researchers published a study in *The Journal of Sexual Medicine* that found birth control pills significantly decrease levels of circulating testosterone, which resulted in a diminished interest in and enjoyment of sex.[23] A lack of sex may be effective, but it's not the type of birth control most women mean to sign up for when they take the pill.

Birth control pills and other hormonal contraceptives also contain a significant dose of synthetic estrogen. When your body realizes that this estrogen-like stuff is pumping through your system, your liver responds by sending out a surge of sex hormone binding globulin (SHBG), the protein you read about earlier that snaps up excess sex hormones so they don't wreak havoc on your body.

The problem is that SHBG has no way of knowing that it's responding specifically to an excess of estrogen. Once released, it indiscriminately snaps up all the estrogen, testosterone, and dihydrotestosterone

(DHT, another important sex hormone) it can hang on to. Your levels of all these sex hormones will drop considerably, but your hormonal contraceptive will deliver a fresh surge of estrogen every day. If you stop using hormonal contraceptives, your SHBG levels will eventually go back down, but studies show that even just six months on hormonal contraceptives can keep your SHBG levels elevated for as long as six months after stopping.[24]

Look, birth control is obviously a personal choice, but oral contraception can take years off your life. It's powerful to have all the information. Don't panic if you've been on the pill for years. It's entirely possible to fix your hormones, get back in balance, and get younger. My own amazing wife is proof of this.

In 2004, right after I got back from the hiking trip in Tibet, I was driving through Arizona with my father when I got a call from a friend in the anti-aging and autism research communities. "Dave, you have to come with me to this conference," she said. She was at the American Academy of Anti-Aging Medicine main event, which to me is like the Olympics. I'd always wanted to go to that conference, and since it was being held in Las Vegas, I happened to be close by.

A few hours later, I walked into my friend's hotel room and met Dr. Dietrich Klinghardt, a renowned expert in Lyme disease, and a beautiful Swedish ER physician named Lana. I was still wearing hiking clothes because I'd come straight from New Mexico, and she was wearing hiking clothes because she'd come straight from visiting the Grand Canyon. We decided to go for a hike, and we've been together ever since.

When I met Lana, she was rail thin, she always felt cold, and she had been diagnosed as infertile because she suffered from polycystic ovarian syndrome (PCOS), a hormonal disorder that leads to enlarged ovaries. She was also thirty-seven, so as we became more serious and discussed starting a family, we knew that getting her hormones to a place where she could successfully have children could be a challenge. But we wanted to take on that challenge together.

First, we worked on improving her diet. We removed the soy milk that was acting as an estrogen mimic and the flaxseed meal that was high in inflammatory omega-6 fats, and replaced them with healthy saturated fat from egg yolks, coconut and MCT oil, and grass-fed meat. Then we cleaned up her environment, including her personal care products, and worked on managing her stress.

Within a year, Lana gained fifteen pounds that looked great on her, saw huge improvements in her energy levels, felt physically warmer, and her symptoms of PCOS started to diminish. We went on to have our two kids without any medical interventions. We published the research we used to fix her fertility in our first book, *The Better Baby Book*, and Lana now uses that same advice in her private practice to help many women struggling with infertility.

Whether or not you choose to have kids, to become Super Human you want your body to be as fertile as possible because our bodies are designed to get out of the way as soon as we can't reproduce. No matter how old you are, you don't want your hormones telling your body that you're past the age of reproduction. A much better signal from your hormones is that you are young enough to have kids and therefore worth taking up room on this planet.

HORMONE HACKS

The good news is that there are many simple ways to hack your hormones besides hormone replacement therapy. Please keep in mind that everything you already read about not dying will also help you balance your hormones. This includes getting good quality sleep, eating the right foods, and avoiding junk light and other environmental toxins. Before you integrate any hacks, I recommend getting an advanced hormone panel from a functional medicine doctor so you can assess your needs.

EXERCISE

Exercise is a simple testosterone booster, and it's one of the most powerful (not to mention least expensive) anti-aging treatments around. Both men and women experience a sharp increase in testosterone and human growth hormone (HGH) after strength training sessions.[25] But high-intensity interval training (HIIT), which involves pushing yourself to your edge with superintense exercise followed by a brief rest, is even more effective at increasing testosterone and HGH levels in both men and women.[26] It's also a great option if you're short on time or don't want to spend an hour in the gym for every workout.

In addition to boosting hormone levels, both endurance exercise (such as running or cycling for longer periods of time) and HIIT help lengthen telomeres.[27] I prefer HIIT because it's much more efficient, but both options will help you stay younger, longer.

No matter which type of exercise you choose, make sure to give yourself time to recover in between workouts and to get plenty of sleep. Sleep time, duration, and quality of sleep all affect the release of HGH, cortisol, and the hunger hormone leptin.[28] Don't skimp on sleep while engaging in intense exercise! And if you track your sleep and see that you aren't getting enough good quality shut-eye, give yourself more time than usual in between workouts to fully recover.

L-TYROSINE

The thyroid is the main energy thermostat of the body, and the hormones it releases control your metabolism and how your body uses energy. Thyroid function often declines as you get older, and with that decline comes a decreased production of thyroid hormones. This is becoming increasingly common in young people as well. In fact, I've seen people as young as twenty who were suffering from the symptoms of premature aging because their thyroids weren't working very well. If your thyroid gland isn't functioning optimally and is failing to produce adequate thyroid hormone, you will feel sluggish and tired and *old*. People with low thyroid are also at a much higher risk of developing heart disease, and women with low thyroid are at an increased risk of fertility issues.

If you're cold all the time and your skin is super dry, I recommend getting a comprehensive thyroid panel from a functional medicine doctor. If those tests show that your thyroid is sluggish or that you have problems converting one type of thyroid hormone into another, an amino acid called L-tyrosine can help upregulate your thyroid function a little bit so you naturally produce more thyroid hormones.

L-tyrosine is also a precursor to dopamine, epinephrine, and nor-epinephrine, three crucial neurotransmitters for focus and mood. Supplementing with it can improve cognitive function while under pressure.[29] In fact, the military has experimented with using L-tyrosine to help soldiers in combat. You can get some L-tyrosine in your diet, specifically in pork, lamb, beef, and fish, but supplementing with its purified form drives your body to produce more beneficial neurotransmitters. When you get L-tyrosine from dietary sources, it comes with other amino acids that the body uses for protein synthesis.

I recommend supplementing with 500 to 1,000 mg of L-tyrosine per day on an empty stomach in the morning. If your thyroid is low or even "low normal" and you have symptoms, ask your functional medicine doctor about trying a small amount of bioidentical thyroid hormone. Even borderline low thyroid is associated with atherosclerosis and raises LDL cholesterol,[30] while adequate thyroid hormones help you stay energetic and lean as you age.

If none of the above works, or if you want to be extremely proactive in your anti-aging protocol, it may be worth considering hormone replacement therapy (HRT). If you go this route, it's critical to work with an anti-aging specialist or functional medicine doctor. You can even google to find one who is trained in the Wiley Protocol.

Bottom Line

Want to age backward? Do these things right now:

• Stop eating sugar, soy, excess omega-6 fats, and refined carbs, and replace these foods with additional healthy saturated fat from grass-fed meat, pastured eggs, and energy fats.

• Exercise intensely one to three times per week to boost testosterone levels. Make sure to recover fully in between sessions. Track your sleep to make sure you are fully recovering!

• Consider taking L-tyrosine, vitamin D_3, vitamin K_2, vitamin A, and zinc supplements to achieve healthy hormone levels. If possible, get your vitamin D and zinc levels tested first to see if yours are low.

• Go through your toiletries and personal care products and get rid of everything containing phthalates and parabens, which mimic hormones in the body and disrupt your natural hormone function.

• If you can, see a functional medicine or anti-aging doctor for a full hormonal workup. If you are deficient in certain hormones and the above advice does not work, explore bioidentical hormone replacement therapy under the care of a trusted physician.

• If you are over forty and have clear signs of low sex hormones, it's probably safe and likely beneficial to try 25 to 50 mg of DHEA without a lab test.

YOUR TEETH ARE A WINDOW TO THE NERVOUS SYSTEM

In 2005, after successfully using laser therapy for all sorts of things I didn't expect, from making my whiplash go away to helping improve my cognitive dysfunction,[1] I decided to attend a training course on how to use the laser in even broader applications. As I sat in a small, cramped room surrounded by about twenty dentists, I stared in awe at our trainer's mouth. He had the most perfect set of straight white teeth I had ever seen. "Why would you drive a Lexus," he asked us, "if you haven't invested at least that much on your teeth, which control your whole nervous system?"

Today we have research to demonstrate the effects of lasers on the central nervous system, but at the time clinicians could only observe the effects of lasers on their patients and speculate about the underlying science. The trainer explained that our front four teeth emerge from the neural crest, a temporary group of cells, when we are embryos, and the back molars are directly connected to the brain. So your teeth can impact inflammation in the entire body, particularly the nervous system.

That trainer used laser therapy on the gums to reduce inflammation in the trigeminal nerve, the largest cranial nerve, which is located along the jaw. The trigeminal nerve controls motor functions in the face and jaw and sends sensory information to the autonomic nervous system, which controls all the bodily functions that happen without our conscious intervention, such as the beating of your heart and the digestion of food.

The little-known field of neurological dentistry recognizes that even micro misalignments in the jaw can cause the trigeminal nerve to send a threat message to your autonomic nervous system, triggering a fight-or-flight response. If your teeth on one side touch before those on the other or your front teeth hit before the back teeth when you bite down, you may be constantly triggering a fight-or-flight response without even realizing it.

What does this have to do with aging? Everything. Remember, a state of fight or flight is literally a physiological state of stress. We already know that chronic stress shortens your telomeres, one of the Seven Pillars of Aging. It also causes your body to consistently release cortisol, the stress hormone, which is highly inflammatory and has its own profound aging effects.

For one thing, excess cortisol triggers the body to store visceral body fat,[2] the internal body fat packed around your abdominal organs. Excess visceral fat is associated with insulin resistance regardless of your weight[3] and inhibits adiponectin, a hormone that regulates your levels of body fat. Too little adiponectin causes your body to pack on excess fat, and studies show that adiponectin levels decrease with increased visceral fat.[4] Both high visceral fat and low adiponectin levels are key indicators of an increased risk of cardiovascular disease.[5] And people with excess visceral fat are more likely to have stiff arteries.[6]

Last but certainly not least, pockets of visceral fat release inflammatory cytokines. But wait, wasn't this visceral body fat caused by the inflammatory stress hormone cortisol? Yes. Excess cortisol causes inflammation, leading to excess visceral fat, which triggers an increase in pro-inflammatory cytokines. This is just one of many reasons it's so important to get your stress response in check to become Super Human. You can meditate and do yoga all day long, but if your bite is still triggering a stress response, you are (and this is the medical term) screwed. Like I was.

The first generation of neurological dentists figured out how to use laser therapy to turn down the fight-or-flight signal sent from the trigeminal nerve to the autonomic nervous system and worked with patients to correct the underlying problem: a misaligned bite. In that hotel room, the trainer melted soft plastic tabs and went around the room to affix them to everyone's rear molars to properly align their bites.

When it was my turn, I could feel my lower jaw relax for perhaps the first time in my life. Until then I had spent my entire life clenching my jaw without knowing it in order to chew. Sure, I experienced occasional tension and pain in my jaw when I would grind my teeth at night. But I couldn't believe that something as simple as a plastic tab that allowed my lower jaw to relax could make such a big difference. And beyond the changes I could feel, the corrected bite took the pressure off my trigeminal nerve, shifting me out of a constant state of fight or flight.

For the next year and a half, I wore those tabs on my back teeth whenever I wasn't eating. In addition to having a more relaxed and better-functioning jaw, I noticed a marked decrease in my overall levels of musculoskeletal pain. If you had told me a few years before that simply relaxing my jaw would help decrease the levels of pain throughout my body, I would not have believed you, and I wouldn't have said it that nicely. But now I am walking proof that proper jaw alignment can help your entire body feel better and become younger. Since the height of your rear molars typically goes down with age, it's a fantastic anti-aging idea to sleep with a bite guard if you clench or grind your teeth.

My hope is that awareness of neurological dentistry will continue to grow until the demand for these services reaches a level that motivates dental schools to begin integrating the necessary training into the classroom.

P IS FOR PAIN

The trigeminal nerve serves as a direct route to the nervous system, and as such it is highly sensitive, with 100 percent more dense pain fibers than any other nerve in your body. Even the slightest dysfunction that puts pressure on the trigeminal nerve leads to elevated levels of a neurotransmitter called substance P, which sends pain signals to the brain. Any time you experience a psychological or physiological stressor, sensory nerves release substance P, which travels directly to the brain, telling it you are in pain. Substance P is a primordial pain-signaling molecule, and a rise in substance P always causes inflammation.

This is an important survival mechanism. In order to live, you have to know when you are in pain! Otherwise, what's to stop you from sticking your hand into a fire or hanging around to see what happens next when you are being attacked by a bear? But like all survival mechanisms, it is a double-edged sword. You experience stress all the time when you are not in immediate danger. For example, when you bite down and your top teeth on one side of your mouth touch down first, this stimulates the release of substance P, which tells your brain that you are under threat and triggers the release of inflammatory cytokines.[7]

It turns out that substance P plays an important role in many health issues we associate with normal aging but are really inflammatory conditions. Patients with asthma are often hypersensitive to substance P, meaning their bodies produce excess inflammatory cytokines when substance P is released. People with eczema and psoriasis also have high levels of substance P.[8] Elevated levels of substance P are found in the colons of patients with inflammatory bowel disease.[9] And studies have shown that cancerous tumors overexpress substance P.[10]

In addition to causing inflammation and playing a role in these diseases, substance P opens up cell membranes, making them less efficient and more vulnerable to toxins that can enter the cell and cause direct damage. You can't detox effectively when your substance P levels are too high. And perhaps most frightening, substance P plays a role in activating stem cells in the body.[11] This means if your substance P levels are out of whack, you cannot efficiently replace cells that die, and the tissues throughout your body will begin to waste away. Yes, like an old person.

Obviously, you must get your substance P levels under control if you want to live a long time without suffering with pain and disease. And the best way to do it is by taking pressure off the trigeminal nerve, which means fixing your bite.

ALIGN YOUR JAW WITH YOUR LIFE

Several years after I attended the laser training session where I learned to use plastic tabs to align my bite, I invited a neurological dentist named Dr. Dwight Jennings to speak at SVHI about aging and jaw alignment. After I heard his presentation, I made an appointment to see him right away.

After he examined me, Dr. Jennings explained that I had a small upper palate and a jacked-up (again, that's the technical term) lower jaw. He built a custom appliance to align my bite that gave me a chin and a square jaw (neither of which I had before), without surgery. Dr. Jennings's work took about two years to complete. First, he made a metal appliance that crossed my teeth in the back and basically pushed out my upper jaw. I wore this appliance whenever I wasn't eating. It affected my speech a little bit in the beginning, but I got used to it quickly, and it wasn't visible. It spread out my lower jaw, making space so it could rest in its natural position. To align my bite once this was complete, we had to widen my upper jaw to accommodate for the wider part of my lower jaw, which was now farther forward. Finally, Dr. Jennings made me denture-like clip-ins that I wore all day long, even when I was eating. At night I wore a bite guard, which I still use to this day to make sure my jaw stays aligned when I sleep. This alone has helped improve my sleep quality and the overall alignment of my body.

Bringing the jaw forward moves the tongue forward as well, opening up your airway. When your airway is restricted, you can snore or suffer from sleep apnea, a sleep disorder in which breathing repeatedly starts and stops. As you read earlier, this common condition can take years off your life. I can feel myself breathing better now that my jaw is aligned, and I know for a fact that I sleep better because I track my sleep!

Beyond sleep apnea, the trigeminal nerve is involved with the brain's reticular activating system, the part of your brain stem that keeps you awake. When there is too much trigeminal disturbance, the brain won't shut down. In this state, you are physically incapable of falling sleep. This is why many people with bite alignment issues suffer from sleep disturbances. But this can—and should—be fixed.

You already know how important sleep is for your longevity, and pressure on the trigeminal nerve can wreck your sleep. Generally our teeth touch for less than five minutes a day when we are eating. But when your bite is misaligned, your jaw is always on guard, trying to keep you from banging your teeth into one another. This puts pressure on the trigeminal nerve and keeps substance P flowing and inflammation building. And because the trigeminal nerve plays such a large role in motor function, it can even cause movement disorders such as torticollis, a condition in which the neck muscles contract, causing the head to twist to one side, and scoliosis, abnormal curvature of the spine.

I hope in the future we will see more dentists studying neurology and fixing jaw alignment to impact spinal health. Imagine if a simple bite guard could help straighten the spine of a child with scoliosis! Before my bite was aligned, you could see the dysfunction in the way I stood and walked, because the trigeminal nerve controls movement throughout the body.

Too many conventional dentists lack this insight. They focus on making the teeth look good rather than bite alignment. I've found that most dentists have no problem throwing off your bite by even a tenth of a millimeter when they fill your teeth because they simply are not aware of how this can impact the trigeminal nerve and therefore substance P levels. Even a tiny misalignment like this can set off a (now sadly shortened) lifetime of inflammation.

Too many dentists also don't realize what proper bite alignment should look like. The standard of care is for the front teeth to hit a bit in front of the bottom teeth instead of end on end. But this is technically incorrect. Dr. Jennings explains that when you bite down, your teeth should all touch at the exact same time—even your front teeth. *This* is proper bite alignment, and it is all too rare.

According to Dr. Jennings, the vast majority of humans these days are moderately compromised when it comes to their alignment, and that's why we have such frequent headaches and other neurological and musculoskeletal problems. He claims that even the high incidence of ear infections in infants can be explained in part by orthopedic jaw defects. If you've been lucky enough to avoid chronic headaches, spinal issues, or other pain conditions, jaw dysfunction can still lead to an inflammatory propensity in the body.

One of the most common and well-known conditions to stem from improper jaw alignment is temporomandibular joint (TMJ) dysfunction, which causes pain in the jaw joint. Most dentists think it's okay for the temporomandibular joints to move back and forth. But according to Dr. Jennings, just because this is the only joint that *can* move back and forth, that doesn't mean it should.

For example, when you have an overbite, you subconsciously have to slide your lower jaw forward to bite things off and to control your airflow and speaking. As a result, you create a sense of hypermobility in this joint, which in return impacts the trigeminal system and leads to elevated levels of substance P. And worse, TMJ dysfunction is often correlated with pain in other joints, likely stemming from the inflammation caused by excess substance P. Dr. Jennings estimates that 30 percent of patients with TMJ dysfunction also have significant knee pain like I did. This is no coincidence.

And there's more. The trigeminal nerve also plays a role in modulating blood flow to the brain. In particular, it controls how much blood flows to the prefrontal cortex, the part of the brain that manages the most complex thinking and decision-making. Again, this goes back to survival. When you are facing a threat to your very existence, you want to shut down the prefrontal cortex so you can act without putting too much thought into it. But it's problematic, to put it mildly, if you can't activate your prefrontal cortex because you're constantly shifting into a state of fight or fight as a result of poor jaw alignment.

When I learned about the role of jaw alignment in brain health, I felt like I'd found another cause of my own cognitive dysfunction. As you read earlier, when I had brain imaging done using the SPECT scan, the result showed virtually no activity in the prefrontal cortex, even when I really tried to think. Could a lack of blood flow, mediated by the trigeminal nerve, have played a role in my diminished cognition? I can't tell you for sure, but it's clear to me that my jaw issues were one cause of my premature aging, and that creating proper jaw alignment is perhaps the most overlooked intervention when it comes to longevity.

DENTAL HEALTH HACKS

The good news is that there are several simple ways to correct bite issues, and many are not expensive. Even a store-bought non-custom bite guard will help somewhat to take the pressure off the trigeminal nerve. Of course, if it's in your budget, a custom bite guard will likely work better and more quickly. There are also several at-home hacks you can try to reduce the amount of substance P in your body. Many of these don't even involve going to the dentist. Here are some of my favorites.

CAYENNE PEPPER

The chemical capsaicin that gives hot peppers their spiciness also reduces levels of substance P in the body.[12] This is one reason people use capsaicin to treat pain. Patches with an 8 percent concentration of capsaicin effectively treat pain for up to twelve weeks,[13] and they do this just by reducing levels of substance P. Capsaicin cream for arthritis works by depleting substance P around a joint.

Hearing this, some people go out and eat a lot of hot peppers or cover their meals in cayenne pepper to reduce their pain levels. You can try simply cooking with more cayenne pepper or taking cayenne tablets. The downside of this is that peppers belong to the family of nightshade vegetables. Many people (about 20 percent) are sensitive to nightshades. For these people eating nightshades causes inflammation, which does not help reduce levels of substance P. If you are one of those people, use pure vanilla bean instead! Both capsaicin and vanilla work by interacting with your vanilloid receptors. Pay attention to your body, and if you start experiencing joint pain after eating nightshades, switch to pure vanilla.

OIL PULLING

In the three-thousand-year-old Ayurvedic practice known as oil pulling, on an empty stomach you swish a tablespoon of coconut, sesame, or sunflower oil in your mouth for up to twenty minutes a day. This ancient practice is known to detoxify and clean the mouth and gums, reduce inflammation and halitosis, and make the teeth themselves whiter. While I haven't seen any studies connecting oil pulling to substance P, it makes sense that anything you do to improve your oral health will also reduce your levels of inflammation throughout the body. You read earlier that some dentists use ozone gas to kill harmful bacteria in the mouth that can otherwise set off widespread inflammation. Oil pulling can also help you avoid this inflammation by getting rid of nasty bacteria hiding in your gums.

The idea behind oil pulling is that the oil literally pulls harmful viruses, bacteria, parasites, fungi, and all of their toxic waste products from the mouth, preventing them from seeping into your bloodstream, where they can cause inflammation and suppress immune and overall health.[14] There have been several promising small-scale clinical human studies on oil pulling. Subjects who used oil pulling reported less gum disease and plaque accumulation than those who simply brushed and flossed their teeth.[15]

This happens because during the swishing process, the oil mixes with your saliva, creating a thin liquid that travels between your teeth and gums to places where bacteria hide. There the oil binds to the biofilm, or plaque, on the teeth and reduces the number of bacteria in the mouth. Many of these harmful microorganisms are coated with fat—a lipid bilayer that is attracted to other fats, including the fat of the pulling oils. Bacteria are absorbed into the pulling oil during swishing and removed when you expel the oil from your mouth. Spitting the oil out instead of swallowing it is an important last step to the oil pulling process, as you don't want to reabsorb these toxins.

When it comes to choosing your swishing oil, coconut oil is preferable to other oils like sesame or sunflower because it works as a natural antibacterial, killing disease-causing bacteria, fungi, viruses, and protozoa. The medium-chain fats found in coconut oil are effective in attacking *Streptococcus mutans* bacteria, which causes cavities.[16] Coconut oil is also naturally anti-inflammatory.[17]

Sure, regular oil pulling has worked fine for thousands of years, but as a professional biohacker I set out to make this technique even more effective. The first step was to switch from coconut oil to XCT Oil, which I created. You absorb oils through your mucosa, so if you're going to go through the trouble of sitting with a mouthful of oil for twenty minutes, you might as well use one that is triple distilled for purity and converts to energy more quickly after it's absorbed through your mucosa. (Like coconut oil, it also has powerful antimicrobial properties.) XCT Oil is made with zero solvents ever, and it is triple distilled in a non-oxygen atmosphere so it is completely free of one type of MCT, C6, which irritates mucosal membranes. Considering how porous your mouth is, a daily oil pulling routine with the cleanest and most effective ingredients makes sense.[18]

Next I add a drop of essential oil of oregano to my XCT Oil before oil pulling. Oregano oil is a well-known and powerful antifungal and antimicrobial, and there's even a study of the combined effects of oregano oil and the most rare form of MCT oil (caprylic acid, which is 70 percent of XCT). The combination was proved to reduce bacteria in meat stored over time better than either caprylic acid or oregano oil alone.[19] I'm willing to bet this means it helps reduce harmful bacteria in our bodies, too.

Whether or not you decide to give oil pulling a try with XCT or just plain coconut oil, both Western and functional biodentists agree that an anti-inflammatory diet low in sugar improves your oral health. If you are following the advice in chapter 3, you are already avoiding the majority of foods that are most likely to cause cavities, infections, and unhealthy gums and therefore trigger an increase in substance P and inflammation.

Having healthy teeth is about more than just looking young or even keeping your teeth until you're a hundred and eighty instead of ending up with dentures, though of course you want to do that, too! Your dental health and jaw alignment will greatly determine how fast you will age and how long you will live. That trainer was right: The quality years I gained from fixing my dental issues are worth much more to me than a Lexus—or for that matter, any car in the world.

Bottom Line

Want to age backward? Do these things right now:

• Test your jaw alignment by opening your mouth, relaxing your muscles, and slowly biting down. What hit first? You want your molars on both sides to hit evenly at the same time and for your front teeth to hit very lightly right after the molars. If anything else happened, look into buying a bite guard. You can get a basic one at the drugstore or a custom job from a dentist if you're able to make a bigger investment.

• Clean up your diet and your mouth. Eliminate sugar, try oil pulling, and brush occasionally with activated charcoal to sop up toxins.

• Find a dentist who will help you try transcutaneous electrical nerve stimulation (TENS) or a cold laser if you suffer from TMJ dysfunction or jaw pain for any reason. This will reduce your substance P levels and take pressure off the trigeminal nerve, literally extending your life.

HUMANS ARE WALKING PETRI DISHES

There is no doubt that antibiotic use was a major contributor to my accelerated aging process. As a child and well into my teens, I was on a course of antibiotics almost every month due to chronic strep throat and sinusitis. These drugs dramatically altered my microbiome, the community of trillions of microorganisms living in the gut that include bacteria, fungi, viruses, and other microbes. We are learning more every day about how the microbiome impacts our health—including the aging process—and many cutting-edge doctors and researchers now believe that the microbes in your digestive tract are in fact calling the shots regarding how quickly you age.

A brand-new study shows that as animals (and likely humans) age, the gut bacteria change, and this change in bacterial composition harms your vascular system, making it stiffer. The study revealed that the gut biomes of older mice have more pathogenic inflammatory species. When these pathogenic bacteria ferment protein, they produce three times the amount of a damaging compound called TMAO than other, more beneficial bacteria. Excess TMAO leads to a stiffening of the vascular system and an increased risk of heart disease. When researchers used antibiotics to knock out the gut bacteria in the old mice, their vascular systems magically became less stiff. The researchers concluded that "the fountain of youth may actually lie in the gut."[1]

But bacteria don't just control us—they *are* us. In 2016, scientists from the Weizmann Institute in Israel found there are approximately

39 trillion bacterial cells in the human body.[2] To date, we know of over a thousand unique species of bacteria in the human gut, and they are not just idle passengers. These bacteria digest your food, keep your immune system humming along, protect your intestines from infections, remove environmental toxins, and produce essential vitamins and chemicals to communicate with the rest of your body. As you read earlier, the mitochondria that power your cells evolved from bacteria, too.

DIVERSITY IN THE GUT

Though all human microbiomes contain roughly the same thousand species of bacteria, the exact makeup of each person's individual microbiome is unique. Your microbiome will look very different from mine. You will have a higher percentage of some species and less of others. However, there are specific hallmarks of a young, healthy, and well-functioning gut. For instance, they contain specific combinations of microbes and, most important, a diverse mix.[3] As you get older, this composition shifts in predictable and unfortunate ways unless you do something to stop it. In fact, researchers at Insilico Medicine, a biotechnology company, can now predict a person's age within four years based solely on their gut bacteria composition.

You can think of the gut as an ecosystem similar to living soil, which is a complex mix of many different bacteria and fungi that work together to make the soil fertile. Since humans don't have roots like plants do, we carry our soil around inside of us. Without the right mix of germs making fertile soil, a plant will die. And we are no different. If our balance of microorganisms is off, we age rapidly, develop disease, and die.

This mix of germs must contain plenty of good bacteria and some bad ones, too. Not every species of bacteria in a healthy gut is beneficial. Even the most Super Human among us has some bad bacteria and even parasites in his or her gut. But if you are young and healthy, your diverse species of good bacteria will overpower the bad ones. The idea isn't to get rid of harmful bacteria completely, but rather to strike a balance between the good bacteria that help you become

Super Human and the inevitable bad bacteria that can cause disease and aging when they overgrow.

There is evidence that having some, but not too many, of these "bad" bacteria and parasites in the mix actually helps keep the gut in balance. We evolved along with these parasites, so our immune system functions better when they are present. In 2005 I read the first study showing that taking certain parasites that cannot reproduce in the human body can alter your immune response and actually *reduce* inflammation. Supplementing with these parasites (which is called helminth therapy) enhances the function of regulatory T cells (Tregs), the immune cells that modulate the immune system and prevent autoimmune disease.[4]

As soon as I read that 2005 study, I ordered pig whipworm eggs from Thailand. They were shockingly expensive, and my friends thought I was nuts, but I was desperate to heal my gut. In all honesty, I didn't feel much of a difference after swallowing those eggs. It's possible that it would have taken more than one dose for those little guys to have an impact, but I couldn't afford to take them regularly at $600 a dose.

Ten years later, after I had grown metabolically younger, I tried helminth therapy again, this time with rat tapeworm larvae. I felt a reduction in inflammation and an improvement in my GI function right away. I've been hacking my inflammation for so long that I can immediately tell how inflamed I am when I wake up in the morning. I can see it in my love handles (or lack thereof) and feel it in my brain. And those little tapeworm larvae definitely dialed down my systemic inflammation. I took them every two to four weeks until it got too difficult to coordinate with my travel, and I plan to take a course of them every six months for good measure. I would certainly continue more frequently if I were suffering from one of the conditions that helminth therapy is proved to treat, such as multiple sclerosis or inflammatory bowel disease.[5]

Thankfully, most of us don't need helminth therapy, since having the right mix of gut bacteria will help prevent inflammation. But over the past several decades, our beneficial gut microbes have taken a hit—we've collectively damaged our microbiomes through an overreliance on antibiotics, antibacterial soaps, hand sanitizers, and the insecticides that we spray on our food.[6] Since gut bacteria have a direct impact on inflammation levels and the immune system, damage to our gut biomes makes us more susceptible to autoimmune disease, which you read earlier is when the immune system attacks healthy tissues in the body.

It's no coincidence, then, that autoimmunity and inflammation have been on the rise in older people for the past several years. In fact, many of the diseases we associate with normal aging are actually underpinned by autoimmune disorders. Approximately 50 million Americans—that's 20 percent of the population—are suffering from an autoimmune disease right now. The standard of care is for autoimmune patients to take drugs that suppress the immune system and leave them unable to fight off everyday viruses and infections. But it is possible to turn autoimmunity around by reducing inflammation and healing the gut instead. This one is especially personal for me. As a young man, I suffered from arthritis and Hashimoto's thyroiditis, both of which are autoimmune diseases. Today I have no symptoms of either condition, and that's without taking any immune-suppressing drugs.

What would you think if you heard that there was a species of probiotic that doubled life-span in mice? There is good science behind the idea of supplementing with specific bacteria to enhance longevity. The specific species with those magic powers works because the bacteria produce spermidine. (Yes, researchers originally isolated spermidine from semen—hence the name. Bear with me on this one.) In one study, when mice were supplemented with the bacteria that produce spermidine, it was shown to *double* their life-span.[7]

You can supplement with spermidine directly, but fair warning—it smells and tastes exactly as you'd expect given from whence it was discovered. As research for this book, I ordered some, held my nose, and swallowed it. But it's easier to get spermidine from the same source as the mice in that study, and that's the species of bacteria that produces it. The more of these bacteria you have in your gut, the more spermidine you produce. I imported this probiotic, which is called LKM512, from Japan, and now I have friendly bacteria in my gut producing plenty of spermidine for me. No more bad-tasting spermidine to swallow, and there's a good chance I'll live longer as a result!

If there's one thing I hope you take from this book, it's the importance of nurturing a healthy and diverse gut biome. But the best way to do that is probably not what you think . . .

GERMS ARE YOUR FIRST GIFTS FROM MOM

Your gut bacteria have been with you for a long time. When you were born, you got your first dose of microbes as you passed through your mother's birth canal. And the way you were born affects the makeup of your microbiome. Studies have shown that children delivered vaginally have a gut biome that is similar to the mother's gut biome. Children who are delivered via C-section have a gut biome that is more similar to the mother's *skin* biome.[8] While the skin is its own diverse ecosystem, the species of bacteria that dominate the skin are typically different from the ones that are most prevalent in the gut.

We are only beginning to understand how the way a baby enters the world might affect its long-term health. Obviously, not every woman is able to deliver vaginally, but one way to counter the potential changes to the microbiome caused by a C-section delivery is to swab the newborn baby with microbes from the mother's vagina. There is some debate about how effective this practice is, but it's a relatively risk-free way of exposing a baby to the bacteria he or she would have otherwise encountered in the birth canal. If you're reading this, you can't change what bacteria you got . . . but if you are planning a family, you can use this knowledge to give your kids a Super Human start.

The baby's first source of nutrition also plays a significant role in shaping his or her early gut biome. Breast milk contains up to six hundred different species of bacteria that help promote bacterial diversity in a child's gut.[9] Unfortunately, formula does not contain these bacteria. And studies show that formula-fed infants tend to have less diverse gut biomes than their breastfed peers, along with an overrepresentation of *Clostridium difficile*, potentially harmful bacteria.[10]

The baby's microbiome continues to evolve during the first few years of its life to support the transition from drinking formula or breast milk to eating solid foods. Studies that have examined stool samples of infants reveal that their microbiomes initially comprise the exact species of bacteria that can best utilize the lactate in the mother's milk. A few months later, the gut shifts to have a higher percentage of bacteria that can metabolize energy from solid foods. In other words, the baby's gut primes itself in preparation to begin digesting solids. Then as the baby eats more and more foods, the bacteria associated with digestion of carbohydrates and vitamins from solid foods proliferate even more.[11] When children are around the age of three, the microbiome stabilizes and becomes similar to an adult microbiome.

The exact makeup of this early microbiome has a huge impact on how the child's immune system develops, and therefore can affect his or her health and longevity much later in life. Specifically, certain species of gut bacteria produce short-chain fatty acids that play an important role in the proliferation and differentiation of immune cells, including T cells and B cells, which produce essential antibodies.[12] As such, the period between birth and the age of three is a critical window in which to establish a healthy microbiome and build a strong immune system. A surplus of bad bacteria and/or a lack of diversity may lead to issues such as autoimmunity, allergies, and asthma.[13]

Finally, exposure to antibiotics in infancy also impacts the health of the microbiome. Antibiotics wipe out both dangerous and beneficial strains of bacteria, causing a reduction in microbial diversity. Studies have shown that early antibiotic use can increase a child's lifetime risk of developing asthma, eczema, and type 1 diabetes, which are all diseases of the immune system.[14] If you're over thirty-five, you lived through the time when doctors handed out antibiotics like candy, and it's likely impacting how you age.

There is no doubt that the makeup of your microbiome during infancy impacts your likelihood of developing specific diseases later in life. In fact, researchers have been able to pinpoint some of the bacterial strains that are crucial for optimal health. For example, infants who develop asthma often have a low abundance of the beneficial bacteria *Bifidobacterium*, *Akkermansia*, and *Faecalibacterium* and a high relative abundance of fungi like *Candida* and harmful bacteria that release inflammatory metabolites. As you know, those inflammatory compounds are at the root of all Seven Pillars of Aging. So what would happen if you had harmful bacteria in your gut releasing them from the time you were an infant? You would age prematurely, that's what.

Of course, there are plenty of things you can do to wreck (or heal—we'll get to that in a minute) your gut biome long after the age of three. All hope is not lost, even if you had a less-than-ideal start, like I did! It's never too late to take control of your gut bacteria. But it's fascinating to know that what happens during the first minutes and years of your life actually plants the seeds for how you'll age decades later.

PROBIOTICS CAN MAKE YOUR GUT WORSE

Given all the bad things that were going on in my gut years ago, I was desperate to fix my microbiome. So desperate, in fact, that in 1998 I ordered a special electrical stimulating pill from Russia. It contained a battery connected to electrodes that was small enough to swallow. As it went through my gut, it used electricity to stimulate the muscles in my intestines. As you'd expect, it felt weird, but it got worse when the pill lodged near a nerve on my left leg. For an entire afternoon, my leg twitched uncontrollably every five seconds when the unit fired.

Needless to say, that wasn't something I wanted to repeat. So I gave up on swallowing weird medical devices from Russia and went back to the gut intervention that most people try first—probiotics. After all, we are bombarded with marketing messages about the health benefits of probiotics, so they must be just the things to fix our gut problems, right? Not so fast.

Unfortunately, many of the probiotics on the market contain an abundance of histamines. When you hear the word *histamine*, you probably think of allergies—we pop an antihistamine pill to block the chemicals our bodies create during an allergic response. But certain bacteria also create histamines through a fermentation process. In addition to playing a role in immune system response (hence the itchiness and sneezing of an allergy attack), histamines act as neurotransmitters, communicating messages to the brain. We need some histamine in our bodies, but an excess of histamine can lead to a condition called histamine intolerance, which causes migraines, sinus issues, and the type of widespread inflammation that is so incredibly aging.

There are two main causes of histamine intolerance: an overpopulation of the bacteria that produce histamines and too little of an enzyme that breaks them down, which is called diamine oxidase (DAO). If you are histamine intolerant, the worst thing you can eat is a food containing histamines. Yet when we are concerned about our guts, many of us turn to the very foods that contain the greatest amounts of histamines, such as fermented foods like yogurt, sauerkraut, or kombucha. Since bacteria create histamines as a part of the fermentation process, fermented foods often (but not always) contain plenty of histamines.

When you're looking to heal your gut, it's important to be aware of which bacteria produce histamines, which degrade them, and which don't affect them at all. Too many people hear about the importance of gut bacteria and start blindly swallowing probiotics, but this is a really bad idea. Remember, everyone's microbiome is unique, and the exact makeup of bacteria in your gut will greatly affect how you age. So you want to make sure you are supplementing with the right species of bacteria *for you*.

The other problem with taking generic probiotics to heal the gut is that if you have an overgrowth of fungus such as *Candida*, probiotics alone aren't enough to fix it. Probiotics are species of bacteria, not fungus. If you have a fungus issue and take probiotics, the fungus will just fight with the probiotics. I was eventually able to get rid of my own *candida* issue by taking antifungal medication for sixty days straight. Incidentally, you can't "starve" *Candida* fungus by going into ketosis because *Candida* can live just fine on sugar *or* ketones. If you know you have an issue with *Candida*, it's vital that you work with a functional medicine doctor to fix it in order for your anti-aging program to succeed. If you don't wipe out the fungus, your beneficial microbes will never have a chance to do their good work.

I know all of this now as the result of trial and error and lots of research. But back when I was first attempting to heal my gut, I made the common mistake of popping probiotics, thinking they would help. I decided to add a prebiotic, or food for probiotics, called fructooligosaccharide to my Bulletproof Coffee in the morning and take a probiotic at the same time. Unbeknownst to me, the probiotic I took contained *Lactobacillus casei*, a histamine-producing species of bacteria. I gained ten pounds in seven days, with noticeable inflammation in my gut. After I stopped the probiotics, it took only seven days for me to lose that weight. It wasn't fat; it was inflammation from the wrong probiotic.

So how do you know which probiotics to take, if you choose to take them at all? If you don't think you have a histamine issue, I recommend probiotics that are neutral, such as *Streptococcus thermophilus* and *Lactobacillus rhamnosus*. To repair an unhealthy gut and decrease histamine intolerance, work on minimizing histamine-producing bacteria and maximizing histamine-degrading bacteria (more on this below). Histamine-producing bacteria include *Lactobacillus casei*, *Lactobacillus reuteri*, and *Lactobacillus delbrueckii* subsp. *bulgaricus*. These are found in most yogurts and fermented foods including sauerkraut, some kombucha, pickles, fermented soy products, soy sauce, fish sauce, buttermilk, kefir, mature cheese, red wine, breads made with yeast, and processed, smoked, and fermented meat. If these foods agree with you, that's great, but pay attention to how you feel after eating them. If you think you might have a histamine issue, avoid these high-histamine products as well as some of the probiotics they contain.

It's possible to take supplements that contain histamine-degrading bacteria such as *Bifidobacterium infantis*, *Bifidobacterium longum*, and *Lactobacillus plantarum*. But if you don't have the right food for those bacteria present in your gut, you're wasting money on probiotics because they'll die before they can help you. Even if you never take probiotics, beneficial bacteria will often naturally grow in your gut when the right fuel is present. Invest in those before probiotics that may not even make it through your stomach. The trick is to focus on eating the foods that help good bacteria grow and reproduce: prebiotic fiber and resistant starch.

PREBIOTIC FIBER

Prebiotic fiber is what it sounds like—the thing that comes *before* probiotics. Simply put, it's what the good bacteria in your gut like to eat. When they feed on prebiotics, these bacteria produce short-chain fatty acids like butyrate, which strengthens your brain[15] and your gut.[16] You can get prebiotics from vegetables that are rich in soluble fiber like sweet potatoes, Brussels sprouts, and asparagus. There is also a little prebiotic fiber in coffee and chocolate, but the best way to live a long time is to start eating a lot more vegetables . . . and maybe add some extra prebiotics, too.

A 2019 review in *The Lancet* demonstrated that eating prebiotic fiber dramatically reduces your risk of developing the Four Killers.[17] In the study, people who ate the most prebiotic fiber had a 15 to 30 percent decreased risk of cardiac-related death and death from any cause, a 16 to 24 percent reduced risk of stroke, and a 19 percent reduction in type 2 diabetes[18] and colorectal and breast cancer.[19] That's a major reduction in three of the Four Killers![20] For the fourth killer, Alzheimer's, we know that prebiotic fiber reduces intestinal and brain inflammation, with a corresponding reduction in inflammation of the immune cells of the brain called microglia.

In another study,[21] researchers gave type 2 diabetics either a 10- or 20-gram dose of prebiotic fiber daily for a month. They saw reductions in insulin resistance, waist/hip measurements, and LDL cholesterol, but the most important anti-aging lab change was in something called glycated albumin. This is a direct measure of the damage sugar is causing as it cross-links the protein in your cells. Another study showed prebiotic fiber works the same way on nondiabetics, too.[22]

The importance of eating enough prebiotic fiber is one of the reasons you hear the bad advice to eat plenty of grains, legumes, and beans. These foods do contain prebiotic fiber, which does great things for your metabolism. Unfortunately, as I highlighted in *The Bulletproof Diet*, they also contain plant defense compounds called lectins, which damage your gut lining and cause inflammation and autoimmune conditions.[23] Those legumes and whole grains are beneficial for balancing blood sugar, but they trash your long-term health by breaking your gut and, as a result, your immune system.[24] Even if you think you can handle whole grains just fine, the evidence is in that a compound found in grains called agglutinins, or WGA, impairs the integrity of your intestinal lining, allowing small molecules to pass through into your blood. You simply won't live as long as you want if you eat grains and legumes. However, you also won't live as long as you want if you don't eat enough digestive fiber. This has been a dilemma for thousands of years.

Technology allows us to have the best of both worlds. Today the best way to feed your gut bacteria is to add at least 10 to 30 grams of prebiotic fiber powder to your diet and eat lots of vegetables. Over the last eighteen months, I've used 50 grams per day of the Inner Fuel prebiotic I formulated with my team at Bulletproof, often blended into my coffee in the morning. During that time, I went from 14 percent to 10.1 percent body fat. Thanks, prebiotics. Acacia fiber, one of the ingredients, also works well by itself, and is widely available.

The science isn't in yet about exactly how much prebiotic fiber you need to live as long as possible. The government recommends about 14 grams per 1,000 calories, which for most people is about 30 grams per day from all sources. But a study in the Netherlands that followed more than a thousand men for forty years found a 9 percent reduced risk of total death per 10 grams a day of prebiotic fiber.[25] Another study in Israel found a 43 percent reduction in total death in people who ate more than 25 grams of fiber per day compared to those who ate less.[26] In that study, an additional 10 grams per day reduced a man's chance of dying by 12 percent and a woman's chance by 15 percent. Yet another study found that increasing your fiber intake by only 7 grams per day produces a 9 percent reduction in cardiovascular disease.[27]

The overarching point is that the right amount of prebiotic fiber is, to be specific, *more*. If you eat five to ten servings of vegetables a day, you may hit the government-recommended fiber levels. However, with this much evidence supporting the benefits, I eat as many vegetables as I can and add another 50 grams of fiber per day to my diet. This is a major upgrade in my recommendations, and the results are tangible. However, if you suffer from small intestinal bacterial overgrowth (SIBO), a condition in which bacteria that are normally in other parts of the gut overgrow in the small intestine, you may need to go on a short fiber-free diet to kill off those bacteria from where they don't belong, so skip this recommendation for now.

RESISTANT STARCH

You read earlier that one benefit of prebiotic fiber is that it supports your bacteria in producing butyrate. You can also get your bacteria to produce more butyrate by eating resistant starch, a type of starch that acts more like a prebiotic than a typical starch, which the body converts quickly to sugar.[28] Resistant starch gets its name from being "resistant" to digestion, meaning your body cannot break it down. Resistant starch moves through the stomach and small intestine undigested and arrives in the colon intact. There it acts as a prebiotic.

There are four types of resistant starch:

- RS1 is embedded in the coating of seeds, nuts, grains, and legumes, which means it is packaged with lectins that harm your gut even though bacteria like to eat it.

- RS2 is the resistant granules in green bananas and raw potatoes.

- RS3 is a type of resistant starch formed when certain starchy foods, like white potatoes (which are a nightshade and thus harm your gut) and white rice, are cooked and cooled.

- RS4 is the man-made resistant starch. The nutrition label on a manufactured and processed food like a bread or cake might include polydextrin or modified starch. This is RS4. Man-made isn't always a bad thing. One study found that resistant dextrin improved insulin resistance and reduced inflammation in women with type 2 diabetes.[29] Just make sure it's not from a GMO source, or you'll be getting glyphosate with your RS4.

Resistant starch does more than just help you produce butyrate. By feeding your good bacteria, it also helps protect you from the Four Killers. A 2013 study found that mice that fed on resistant starch had a decrease in the number and size of lesions associated with colon cancer. The resistant starch helped kill precancerous cells and reduced the systematic inflammation caused by cancer.[30] It also helps to reduce insulin resistance. Since resistant starch isn't digested, your blood sugar and insulin levels don't rise after you eat it. A 2012 study found that obese men who consumed 15 to 30 grams of resistant starch every day for four weeks showed increased insulin sensitivity compared to a control group that ate no resistant starch.[31] Insulin sensitivity, the opposite of insulin resistance, is incredibly important for your longevity.

Eating resistant starch is also beneficial for weight control. One study found that women who ate pancakes made with resistant starch burned additional body fat after the meal compared to women who ate pancakes without the added resistant starch.[32] This isn't surprising since your good bacteria influence your metabolism and play a critical role in weight management. We've known for years that obese people and lean people have different types of microbes in their guts.[33] Obese people tend to have more of a type of bacteria called Firmicutes and fewer of a type called Bacteroidetes[34] than thin people. This is true even in cases of twins, when one twin is obese and the other isn't.[35]

You can't buy Bacteroidetes as a supplement, but you can get more of them by eating spices and vegetables that contain polyphenols, which are the preferred food source of Bacteroidetes. When you eat a diet high in polyphenols, your Bacteroidetes thrive and reproduce. As a general rule, the more vibrant a vegetable's color, the more polyphenols it contains. Vegetables that are dark green, deep red, purple, orange, and bright yellow all have high polyphenol content. Coffee, tea, dark chocolate, and fresh herbs and spices are exceptional polyphenol sources as well.

Your gut bacteria also impact your weight by producing a hormone called fasting-induced adipose factor (FIAF), which tells the body to stop storing fat and to start burning it instead. The best way to ramp up FIAF production is to starve your bacteria of starch and sugar. When bacteria are "hungry," they make more FIAF, and you burn additional fat. This is yet another reason it's so important to fast occasionally if you want to keep getting better with age.

BACTERIAL FUEL AND THE GUT LINING

As you likely know, the vast majority of your gut bacteria live along your gut lining, which is made of mucus and acts as a barrier to protect your body from the contents of your digestive tract so that dangerous microbes don't leak into your bloodstream. When things are working well, your gut lining allows nutrients in and keeps disease-causing pathogens out.

Your gut microbes themselves help to maintain the integrity of this gut lining. The butyric acid they produce when they eat prebiotic fiber or resistant starch fuels the cells that line the intestinal wall. This keeps it strong and healthy and prevents you from developing leaky gut syndrome, a condition in which microscopic holes form in the gut wall, allowing intestinal contents to "leak" through the barrier and into the bloodstream.

When you have leaky gut, proteins can get into the bloodstream and trigger allergies or even an autoimmune attack. So can bacteria and bacterial neurotoxins called lipopolysaccharides (gut researchers call them LPSs), which definitely don't belong in your bloodstream. Once they leak out, they can impact other organs like the liver, kidneys, and heart, causing widespread inflammation and disease.[36] Leaky gut has been linked to autoimmune disease, type 1 diabetes, inflammatory bowel disease, celiac disease, multiple sclerosis, and asthma, among other killers.[37] Less serious but more common issues caused by leaky gut include acne, rosacea, stomachaches, headaches, and fatigue. In fact, I believe LPSs are a primary cause of inflammation and the aging it creates. You simply must reduce the quantity of lipopolysaccharides your gut makes and the quantity that can get into your bloodstream.

To protect your gut wall, you need to give your good bacteria the food they need to thrive. A 2018 study published in *Cell Host & Microbe*[38] reveals that good gut bacteria—specifically bifidobacteria—rely on fiber as a nutritional source to maintain a healthy gut lining. In the experiment, mice that were fed a low-fiber diet developed leaks in the mucus layer of their gut linings after only three days. The fiber-deprived mice then received a gut bacteria transplant from normally fed rodents, and they regained some of the protective coating necessary for a healthy mucus layer.

When these mice later received a probiotic supplement of bifidobacteria, their mucus layer grew, but it did not repair the gut lining's permeability. But adding a type of prebiotic fiber called inulin to their diets fixed this issue. The researchers concluded that bifidobacteria are crucial for proper functioning of the gut lining, and that—no surprise—bifidobacteria rely on prebiotic fiber to grow and multiply.

This is a big deal. The movement of bacteria and toxins from the gut to the rest of the body is one of the most significant and preventable causes of aging in our modern society. It can prompt or worsen a chronic inflammatory response that causes rapid aging, and it can even lead to mental health issues. This is because gut health and brain health are closely connected—the gut and the brain communicate with each other constantly by sending chemical signals along what is called the gut-brain axis. A groundswell of research in recent years points to a strong link between what's going on in your gut and various mood and behavioral disorders[39] including depression,[40] autism,[41] and even neurodegenerative diseases.

A 2018 study out of Japan found that transferring the fecal bacteria of depressed people to the intestines of rats led to depressed behavior in the rats.[42] And another recent study showed that gut bacteria are in charge of activating certain parts of the brain during times of stress. Researchers analyzed the stool of forty healthy women and then divided the women into two groups based on their gut bacteria composition. They then showed the women negative images while monitoring their brains. It turned out that the dominant bacteria in the women's microbiomes determined which parts of the brain were most active while viewing negative images.[43]

There is clearly more to the gut-brain axis than we currently understand, and I'm excited to keep learning more. We do know that stress directly affects the gut. One study showed that exposing participants to a stressor actually changed the makeup of the microbiome, decreasing the relative abundance of one beneficial species of bacteria while increasing the relative abundance of pathogenic bacteria.[44] These changes make people who experience significant stress more likely to develop major gastrointestinal disorders, including inflammatory bowel disease (IBD), irritable bowel syndrome (IBS), and gastroesophageal reflux disease (GERD).[45] Have you ever thought that stress was affecting your gut? You were right.

So your gut has the power to change your brain, and your brain has the power to change your gut—and how you age along with it. For millennia we've thought of the inner workings of the gut as an unknowable mystery, but thanks to new technology and computing power, it doesn't have to be that way.

TRACK YOUR GUT LIKE YOU TRACK YOUR SLEEP

By far the best way to figure out how and whether you need to heal your gut is to find out exactly what's going on in there. Right now the most effective way to do this is through Viome, a company that uses technology that was developed by the U.S. defense department to detect biological warfare to analyze what's going on in your gut. After you send a stool sample swab to Viome, they are able not only to identify every organism present but also to assess how active they each are by looking at what sort of beneficial or harmful compounds each type of bacteria is producing.

While identifying the microorganisms in the gut is important, it's even more helpful to understand their function. The microbes in the gut produce thousands of chemicals that affect how quickly you will age. By analyzing the genes that microbes express, Viome can identify which of these chemicals they produce and can determine their role in your body's ecosystem.

Every living organism produces RNA molecules from their DNA. Viome sequences all the RNA in the samples it receives; in this way your gut's living microorganisms (including bacteria, viruses, bacteriophages, archaea, fungi, yeast, parasites, and more) can be identified and quantified at the species and strain level. The end result is a higher resolution view of your gut microbiome than has ever been available.

Viome then feeds this information through its artificial intelligence technology and sends you a report that tells you which foods will feed the good gut bacteria you want to foster and which foods are causing an imbalance of gut bacteria. Viome's report lets you fine-tune your gut microbiome function to minimize your production of harmful metabolites and maximize your production of beneficial ones. If this isn't a powerful anti-aging strategy, I don't know what is.

Full disclosure: I joined the advisory board of Viome because it's the first technology I've found in twenty years of searching that could actually tell me what was going on in my gut at every level. I think it's a world-changing technology that's using big data to allow us to look inside the black box of the gut.

If it's outside of your budget or you decide not to get customized dietary advice from Viome, the best way for anyone to starve bad bacteria and feed good ones is by cleaning up your diet. This means:

- Don't eat grains, legumes, or nightshade vegetables, all of which lay the groundwork for leaky gut syndrome.

- Quit eating sugar. If you make one change to improve your gut health, make it this. Bad bacteria love sugar and feed off it. Excess sugar is the prime culprit behind small intestinal bacterial overgrowth (SIBO) and *Candida*.

- Never eat industrially-raised animals again, because the antibiotics they receive and the glyphosate in their food will end up in your gut and harm your gut bacteria.

- Feed your gut bacteria a whole lot more prebiotic fiber. Eat a variety of polyphenol-rich vegetables, drink coffee and tea, and add at least 10 grams of prebiotic fiber. I use 50 grams per day of Bulletproof Inner Fuel, but you can also use plain acacia fiber.

- Add medium-chain triglyceride (MCT) oil to your diet. As you read earlier, the saturated fatty acids found in coconut oil have antifungal, antibacterial, and antiviral properties. I recommend Brain Octane Oil because it raises ketones more than generic MCT oil.

- Get more grass-fed collagen protein. Collagen helps your body maintain the gut lining so you can avoid leaky gut and more easily absorb nutrients.[46] Eat collagen-rich foods such as bone broth, and add grass-fed collagen protein powder to your smoothie or Bulletproof Coffee.

My Viome results provided both good news and bad news. On one hand, I am still recovering from antibiotic exposure even though I haven't taken antibiotics in years. Because I've taken so many antibiotics, my bacterial genes show that I am resistant to five different strains. My first Viome test also found increased amounts of human DNA, which sounds like a good thing but actually means that my gut lining is turning over more than it should, causing inflammation. Finally, the test showed average metabolic fitness.

It was these results that helped me discover that I wasn't eating enough vegetables when I traveled, which led me to create a prebiotic blend I could take with me. After I used it for three months, my Viome test results shifted. My inflammation levels are now in the lower 27 percent of the population, and my metabolic fitness went from average to high, in the top 18 percent of the population. In addition, I went from the low end in terms of bacterial diversity (48 species) to the high-normal range (196 species). I'm certain that the changes to my gut are already helping me toward my goal of living to a hundred and eighty.

Bottom Line

Want to age backward? Do these things right now:

• Up your intake of prebiotic fiber, resistant starch, and polyphenols, and cut way back on sugar. This alone will take you a long way toward balancing your gut biome.

• If you have GI issues, cut back on fermented foods, including yogurt, sauerkraut, and kombucha, to see if it helps. You may be histamine sensitive.

• Consider getting your stool tested by Viome to find out what's really going on in your gut. Viome is generously offering a significantly discounted rate to readers at viome.com/superhuman.

• You might also want to consider other at-home lab tests from reputable companies such as EverlyWell, which offers everything from food sensitivity tests to thyroid, inflammation, and other hormone panels. This is valuable information!

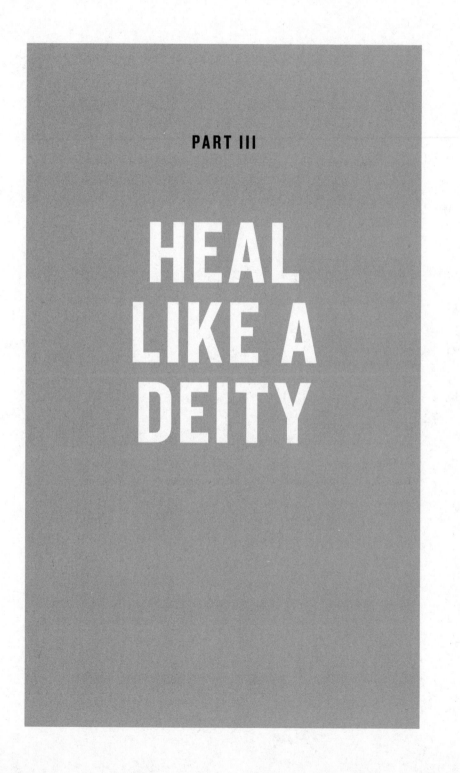

PART III

HEAL LIKE A DEITY

Remember Thog, the caveman who was skeptical of fire, and his friend who embraced that new technology? Perhaps their first glimpse of fire came from a naturally occurring forest fire, but according to Greek mythology, the Titan Prometheus, a champion of knowledge for the human race, gave mankind the gift of fire. To help enable our progress, Prometheus stole fire from the gods and gave it to humans.

Any time you share new knowledge, there will be pushback. In this case, it really pissed off Zeus, the king of the gods. To punish Prometheus, Zeus tied him to a rock and sent an eagle to eat his liver every day. Being an immortal and all, Prometheus regrew his liver every single time. Infinite pain and suffering inflicted by the gods is harsh, but it's a fascinating myth that ties together two major themes of this book: innovation and regeneration. Fortunately, modern anti-aging researchers don't face liver-eating eagles, but they do face a never-ending army of naysayers and regulators seeking to slow the inevitable progress of this most important work.

After I got my biology back to functional, my focus turned to finding a new level of youthfulness by helping my body regenerate—if not literally like Prometheus's liver, then at least like a young person. Thanks to abundant active stem cells and growth factors, young people heal from injuries and everyday slights much more efficiently than most people do as they age. This is one of many reasons older people suffer from aches and pains that keep getting worse and never go away.

To live to a hundred and eighty with all my faculties intact, I have to do whatever it takes to enhance my body's healing mechanisms. Full disclosure: Not all of the things I've tested to improve my ability to regenerate are approved by regulators, and many are not regulated at all because they are too new. Some were also stupidly expensive at first, but the costs are falling rapidly. All interventions carry risk, as does a visit to your doctor. According to Johns Hopkins, medical errors are the third leading cause of death,[1] which means it could rate as one of the Four Killers. I chose to try these interventions because they have good science behind them. I've compared the risks with the potential rewards and consciously decided to take on those risks in exchange for the rewards because it's worth it to me.

Your risk threshold might be lower, and we all have budgets. That's why I've gone to great lengths to include options with less risk and/or more affordability; thus you can access some of the same benefits. But to become Super Human means venturing into uncharted territory, just as Prometheus did and the caveman who first accepted his gift of fire. I'm excited to share the tools I've used to start healing like a teenager and maybe even someday like a deity. It's up to you to decide how far you're willing to go to give death the finger once and for all.

VIRGIN CELLS AND VAMPIRE BLOOD

Through my work in the anti-aging nonprofit field, I followed the research on stem cell treatments for years and desperately wanted to try it myself. In my early thirties, I had old injuries from my childhood that still bothered me, and of course I wanted to find new ways to help my body heal and regenerate. For about ten years, I couldn't find a way to make it happen. At the time, these treatments cost upwards of $150,000 and required a trip overseas. Only professional athletes even considered it.

This is because, unfortunately, most of the cutting-edge research on stem cell treatments has occurred outside the United States. Back in the 1980s, researchers here originally studied stem cells that came from an eight-cell human embryo. This raised a huge controversy from people concerned about experimenting on fetuses. In response, the government shut down much of the research on stem cells in general, even though stem cells are found throughout adult bodies, too. We just didn't know it then.

Let me be clear: None of the treatments I've pursued have anything to do with embryonic stem cells. Current therapies use adult stem cells from the patient's own body or placental or cord tissues left over from healthy pregnancies that we used to incinerate as waste until we learned about their healing powers. Yet only certain stem cell treatments are legal in the United States, and often those have restrictions that make them less effective and a lot more expensive than in other countries. While I appreciate having regulatory agencies looking out for bad actors, we have a basic human right to choose what we want to do with our own bodies whether or not a church, a specific doctor, or a regulatory agency has blessed that decision. My biology, my choice.

CELLULAR FIRST RESPONDERS

So why are stem cells so important? Stem cells are the body's master regeneration cells. We have them in many tissues in our bodies. When they are specialized—like a neuron—they can only turn into other neural cells, which means they are differentiated. For instance, adult immune stem cells aren't supposed to turn into muscle cells or neurons, but they can turn into various different types of immune cells. Of course we've already figured out how to break these biological rules, but more on that later. The most powerful type of stem cells are pluripotent, which means they have unlimited self-renewal capacity, and they can differentiate into any type of cell in your body. Embryonic stem cells (found in days-old embryos, not fetuses) are pluripotent, meaning they can divide and become any other type of cell.

Researchers discovered stem cells in bone marrow sixty years ago, and we've been doing stem cell transplants using marrow to cure some types of cancers and other life-threatening diseases for forty years. Only relatively recently have we learned that there are also dedicated stem cells in your heart and brain. Stem cells are sneaky—they hide without dividing for long periods of time until an injury or the need to grow new cells as you age activates them.

A stem cell's job is to maintain and repair the tissues where it lives. It's like a consultant that rolls up on the scene, assesses the situation, and then acts to make improvements. Whenever you heal from an injury or regenerate tissue, it is a stem-cell-mediated event. As a result, your stem cell function largely determines your overall health and longevity. Your tissues are constantly renewing, and this requires a healthy and robust population of stem cells. But as you age, that reserve of stem cells can become depleted. With fewer stem cells in the reserve, cells that die are not automatically replaced. The body loses the ability to heal itself, and injured tissues deteriorate. This is called stem cell exhaustion, and it's why older people typically don't heal from injuries as quickly as they did when they were young.

In addition, stem cells themselves begin to show signs of age. They don't replace dead cells as efficiently as they did before. This is not merely a superficial problem. When your body cannot heal itself efficiently, the result is nagging pain that ages you even more. You read earlier about how substance P leads to inflammation. Stem cell exhaustion causes pain and increased substance P levels, which leads to inflammation, which causes even more aging!

In your own body, stem cells are greatly concentrated in subcutaneous fat and in bone marrow. A doctor can harvest a little bit of your fat or bone marrow, process out the stem cells, and then reinject them into your body, where they dial down inflammation and promote healing. Of course an aggressive anti-aging strategy must include stem cell treatments. And mine sure has. But stem cells aren't the only type of cell or compound that you can pull out of your body and reinject to promote deity-like levels of youthfulness. Several types of cells and growth factors are more prevalent in your blood when you are young. Boosting these levels through cutting-edge treatments may be one of the most impactful ways to reverse aging.

For years I had been trying to find a very experienced stem cell doctor in the United States. Finally in 2015 a friend introduced me to Dr. Harry Adelson, who was an early adopter of stem cell therapies to treat the pain that too often comes with aging. For many years, his practice largely consisted of farmers, ranchers, oil field workers, and professional rodeo folks. These are people who often have arthritis throughout their entire bodies and need to heal from injuries to continue working or to enjoy retirement. As such, he frequently does large treatments, injecting stem cells into many body areas in a single sitting.

Many stem cell treatments use stem cells from either the fat or the bone marrow of a patient, but Dr. Adelson prefers to use a combination of both. He calls the stem cells from bone marrow the workhorses of the stem cell world. They are not as plentiful as those found in fat, but they are potent and come with lots of beneficial growth factors. Adipose tissue, which stores fat, is rich in stem cells that are more plentiful but have fewer growth factors than those found in bone marrow. By combining both types of stem cells in one treatment, you can get the best of both worlds.

I lay facedown on an exam table, and the doctor smeared numbing cream over the area best described as my love handles. This is where fat most often accumulates and is a rich source of stem cells. Some of the sweetest words I've ever heard were "Don't lose any more fat, Dave—you barely have enough." Then he injected a local anesthetic and performed what was essentially liposuction to remove a few ounces of fat and the stem cells living inside it.

I actually did a Facebook live stream during the procedure, and in selfie mode, I could see everything that was happening behind me. (I made sure to keep my head up and used a strategically placed blanket so my butt wasn't too visible.) It looked like the doctor was tenderizing a steak, but I couldn't feel a thing. Over the course of a few minutes, he removed about a coffee cup's worth of fat.

Then it was time to get more stem cells from my bone marrow. I had heard this might be painful, so I was a tiny bit nervous, but it wasn't actually that bad. After making sure I was completely numb, Dr. Adelson made a small incision at the top of a butt cheek, and then I felt a sense of pressure and heard the *dink, dink* of a hammer hitting up against something hard: my bone. I won't lie, it was uncomfortable and sort of creepy, but I just kept reminding myself how badly I wanted to live well for a long time. It was nowhere near as painful as I'd heard, just really odd because your skeleton is not normally a source of noise and vibration.

It was over quickly, and Dr. Adelson spun my fat and bone marrow in a centrifuge to extract the stem cells. Much to my delight, he commented on the fact that my bone marrow was a rich yellow color (similar to grass-fed butter . . . you are what you eat) and held more stem cells than he normally sees.

I like to think this is thanks to all the work I've done to age backward. Before the stem cell treatment, I spent a lot of time in the Atmospheric Cell Trainer, a piece of technology from Upgrade Labs, the first biohacking facility in the world. The trainer looks like the cockpit of a jet fighter, and you sit in it while the air pressure rapidly changes from sea level to as high as Everest and back. This causes all the cells in the body to grow and shrink, impacting circulation and likely stem cells as well.[1]

Did I have unusually high levels of stem cells because I used the Atmospheric Cell Trainer, because I wrote a book about how to make mitochondria work better and practice those lessons every day, or because I meditate? Who knows? And honestly, who cares as long as the results are there? Though we've made a lot of progress toward understanding the body over the last ten years thanks to computing and sharing information between fields, the body is still a black box. To a biohacker like me, a black box is a system. I don't need to know everything that's going on in there. All I know is that I can put something in the box and get something else out.

While academics and engineers want to take apart the box and understand all the components, hackers say, "Let me know what you find. In the meantime, I'm going to keep changing things until I get what I want out of the box." This is where we are with anti-aging right now, and probably will be for the next fifty years. At some point, someone will figure it all out down to the subatomic particles and flow of electrons. My goal is to stay alive long enough to benefit from that knowledge, and maybe even contribute to it.

Once Dr. Adelson had all the stem cells processed, he began injecting them throughout my body. Using a 3-D X-ray machine to make sure he had exactly the right spot, he focused on injecting stem cells into old injury sites. As a fat nineteen-year-old soccer player, I once dove onto a ball to save a goal and landed on my right shoulder, damaging the rotator cuff. I had also suffered from upper back pain for years.

With so many fresh stem cells, I decided to have some injected into my face to help keep my skin collagen- and elastin-rich and some put into my reproductive organs. While I've never had a problem with erectile dysfunction, this is a common and well-known symptom of aging. Stem cell treatments can help keep things flowing with increased blood flow and nerve response. Hopefully with this help, I won't ever need a little blue pill, even when I'm a hundred and eighty.

Dr. Lana observed the entire treatment. As a physician who is excellent at evaluating medical procedures and who of course wants to grow young with me, she decided to get the same treatment. She, too, had a lot of old injury sites to heal. When she was eight years old, she fell about thirty feet out of a tree and landed on her back. And just two years later, she was playing on the second story of a construction site when a friend accidentally pushed her out the window. She had pain in her neck and limited motion for four decades.

Lana also decided to get the equivalent female sexual health treatment, which included stem cell injections into the clitoris and the upper vaginal walls. Most women know that as they age or have babies, it's common to see a thinning of tissue thanks to cell loss, which causes a decrease in sexual pleasure. The stem cell treatment helped those tissues regenerate. Within days, Lana's neck pain was completely gone, and she could turn her head with full range of motion for the first time in her adult life. My shoulder and back pain went away. It was incredible.

I was so impressed with the results that I recommended Dr. Adelson's treatment to many more family members. One was scheduled to have surgery on a damaged heart valve. Beforehand, he had his own stem cells pulled out and reinjected through an IV for anti-aging purposes and in preparation for the surgery. Soon after, he went to the hospital to have a routine heart scan prior to the surgery, and his doctor told him, "You have no damage in your heart. It's gone. There's no need for the surgery."

Not long after this, my mother fell down, and her glasses cut her face right under her eye. It required eight stitches and left a large, noticeable scar. About three months after the fall, I gifted stem cells to her and my father, who had them introduced intravenously so they could go to sites of inflammation. Very quickly, her scar shrunk to the point that you could barely see it. When a nearly seventy-year-old woman magically heals from what would have been a disfiguring scar, we're getting close to deity-like powers.

FULL-BODY MAKEOVER WITH STEM CELLS

Since I'm always looking to up my game and I wanted to share the experience with you in this book, I recently went back to Dr. Adelson for a treatment he pioneered called the full-body stem cell makeover.

In fact, I was the first person on Earth to receive the highest level, which he calls the six-hand full-body stem cell makeover. For this treatment, I was sedated via IV. This is slightly different from being put under general anesthesia, which uses drugs that tax the brain and liver. The IV sedation is the same kind that is typically used when you get a colonoscopy or during certain dental procedures.

Once again Dr. Adelson removed and then reinjected a combination of stem cells from my bone marrow and adipose tissue. This time, he added exosomes, which are vesicles (fluid-filled sacs) filled with growth factors made by stem cells. These exosomes came from umbilical cord stem cells cultured in a lab. When placed in a stressful culture medium, these stem cells believe their host is under duress, so they manufacture and release vesicles filled with growth factors to help fight off whatever threat is coming.

These exosomes are essentially the active ingredients in stem cells. You can think of them as stem cell juice (yum!) with all of the stem cells strained out. They are responsible for the intercellular communication that triggers the growth of new tissues and new blood vessels, controls inflammation, and fights off infection. When our stem cells age, they lose the ability to manufacture these very exosomes. The lab forces very young, robust stem cells to sprout exosomes. Then they separate the exosomes from the stem cells and discard the cells that contain the other person's genetic material. The exosomes themselves do not contain any genetic material, and their membranes are identical to the membranes of our own stem cells. So researchers believe that our stem cells are able to absorb exosomes, effectively making the stem cells themselves younger. This assumption is based on scientific literature on the use of exosomes for kidney function.

As a side note, some people consider the use of umbilical cord blood controversial, but there is currently a thriving market for both umbilical cord blood and amniotic fluid, which contain growth factors such as exosomes that promote rapid healing. It may seem icky or unethical for women to donate (or even sell) their cord blood or placenta after delivering a child. When you consider that these tissues were previously incinerated but are now helping people heal faster, however, the ethics are clear.

In an ideal world, we'd find some exceptionally strong and vital stem cells from an umbilical cord and then grow them in a lab until we had enough "super stem cells" to treat tens of thousands of patients at a very low cost. But this is illegal in the United States, so for now we are stuck with using only the stem cells present in an umbilical cord, which makes it harder to test for genetic and biological issues. Umbilical cells are routinely tested for seven common diseases, and donors go through extensive interviews to reduce risk, but there is probably a lot more testing that can be done to ensure better results. Because of these restrictions, in the United States the normal dose is 3 million cells, and I have to take it on faith that the cells are alive, strong, and from a healthy person. When I go overseas, I can get a dose of 200 million cells that were strengthened, cultured, and tested more extensively.

The same thing goes for your own stem cells that are stored in a stem cell bank. It's legal to "bank" some of your cells to use at a later date. They become frozen in time, so my forty-year-old stem cells will still be forty when I'm a hundred and twenty, assuming the stem cell bank doesn't have a power outage! In an ideal world, the stem cell bank could strengthen, culture, and grow my cells so I'd have dozens of doses. Some do offer this service, but no doctor here will reinject them into you because the government considers your own cells unlicensed drugs when they are taken out of your body and cultured. Every time I have stem cells taken out, I bank some in case of a future injury or surgery. But if I have them cultured to filter out the weak ones and encourage the strong ones to grow, I'd have to go overseas to put them back in.

I'm really hoping that in the near future we will lift some of these regulations and increase demand for these treatments so they become safer and more accessible for everyone. At this point, I recommend banking your stem cells right away if you can afford it so you'll always have a supply of your own stem cells that are younger than you are. If you get one of the Four Killers or any big injury, those stem cells could save your life or maybe just keep you young.

For the full-body stem cell makeover, I was allowed to use exosomes because they don't contain genetic material, and they are an approved product in the United States. For the procedure, Dr. Marcella Madera, a neurosurgeon and Johns Hopkins–trained spine surgeon, injected the combination of stem cells and exosomes into my entire spinal canal. The goal was to prevent future central stenosis, a narrowing of the spinal canal, which is a form of arthritis. Then she injected it directly into the cerebrospinal fluid so it would cross the blood-brain barrier and regenerate my brain.

Next, Dr. Adelson performed stem cell injections into the posterior column of my entire spine from the base of my skull to my tailbone, into all the major peripheral joints (shoulders, elbows, wrists/hands, hips, knees, ankles/feet) to reduce my chances of osteoarthritis, and into the major tendons to keep them strong and avoid tendonosis or rupture. He also injected Wharton's jelly, the gelatinous substance that insulates the umbilical cord, into all major joints. This Wharton's jelly is made up of bioidentical human-derived hyaluronic acid and chondroitin sulfate, the building blocks of intervertebral discs, joint surfaces, and ligaments. This gives the stem cells the materials they need to grow new connective tissues. Having experienced them as a teenager, I really don't want creaky joints when I'm a hundred!

To top things off, Dr. Amy Killen injected stem cells into my face to improve the health of my skin, into my scalp to improve the thickness of my hair, and into my penis to increase microcirculation and sexual function. I woke up looking a bit like Frankenstein's monster. My face was red and swollen, my hair was spiked, and I was a little sore from the bone marrow aspirate. But after some serious rest for two days I was as good as new—better, actually.

Within sixty days of the treatment, my amount of deep sleep started to skyrocket, and my REM sleep increased dramatically as well. They are now the highest levels since I started recording them. On some nights I am getting 2 to 3 hours of REM sleep and 1.5 to 2 hours of deep sleep, even if I sleep for only 6 hours total. As you read earlier, these are the numbers of a much younger person.

Over the next couple of months I noticed that my resilience greatly increased. It's always been a challenge for me to keep my brain working well when I'm exposed to a lot of junk light, especially when I'm traveling heavily with fourteen-hour nonstop days. One of the first things to happen when I'm at capacity is that lights hurt my eyes. A few days after the procedure, my brain resilience was higher—I now experience less visual stress around crappy lighting.

A month after the full-body makeover, I had to fly back and forth from the West to East Coast four times within just two weeks, which is a brutal amount of travel. At the same time, I was dealing with an exceptionally stressful family situation, a crisis at one of my portfolio companies, and the looming deadline for the book you are reading now. It was the highest amount of combined stress I'd faced in fifteen years.

I meditate, I teach CEOs how to meditate with electrodes on their head, and I have designed nootropics to reduce anxiety and increase cognitive performance. Yet I'm still (mostly) human, and this level of stress should have left me depleted. Instead, I had far more energy than I expected. It was as if I'd reached another level of performance and resilience that I didn't know was there before—aka I was younger.

At around the same time, I had the opportunity to have Russian Stick Bodywork done from the Somatic Training Network. This is a form of very intense bodywork performed as a part of the Systema martial arts practice. It is in part based on a shamanic lineage from Siberia, and it's the most hard-core thing I'd ever experienced. The founder of the Somatic Training Network, Dan Sykes, works with old Russian masters who have decades of experience working with elite military units. They push an implement resembling a drumstick deep into your muscle beds, walk on you, and then (you can't make this up) beat you a few times with a whiplike device to "wake up" local parts of your nervous system.

When he was done, the wizened master of this technique looked dumbfounded and said to Dan in his heavy Russian accent, "His muscles are like sand. There is no resistance. Very advanced. I have never seen this." I like to imagine that's what Prometheus's muscles were like, too.

GET MORE STEM CELLS WITHOUT INJECTIONS

If stem cell treatments are not in the cards for you right now, never fear—there are other ways to stimulate stem cell production and availability. One of the most remarkably simple and inexpensive methods is to ingest the mineral boron, which is commonly found in the household cleaner Borax. It sounds crazy, but there have been anecdotal reports since the 1960s that boron can help ease arthritis pain,[2] and we are finally beginning to understand why.

When researchers added boron to frozen stem cells, it increased their viability and ability to form bone and cartilage. The researchers concluded that it helped the stem cells withstand the stress of being frozen.[3] Perhaps this is why studies consistently show that boron helps reduce joint inflammation[4]—the boron strengthens stem cell viability, so senescent cells in the joints are replaced with healthy cells, reducing pain and inflammation. Boron may also keep the stem cells stored in your fat from turning into fat cells, saving them for other uses in the body.[5] I haven't had joint pain since my stem cell makeover, but I supplement with calcium fructoborate, a form of boron that is widely available as a supplement, as extra insurance . . .

Here are some additional ways you can increase your stem cell production, no injections required:

- Fasting for twenty-four hours or more can double the regenerative abilities of stem cells, according to MIT research.[6]

- Short-term calorie restriction also increases stem cell activity, even if you're not fully fasting.[7]

- Drugs called PPAR agonists also appear to increase stem cell performance. The most common is the diabetes drug Actos (pioglitazone). An off-label use may be useful for stem cell enhancement. Natural compounds that may work on the same pathway are sesamin (an extract of sesame) and fish oil.

- Cut sugar from your diet and reverse insulin resistance. In culture, stem cells "exhibit greater self-renewal and anti-senescence abilities" with restricted sugar. The same is likely true for you![8]

- Lift heavy things. Research shows that even a single weightlifting session improves stem cell activation.[9]

- Turmeric supplements cause brain stem cells to grow both in live animals and in lab cultures.[10]

- Resveratrol (a type of polyphenol) supplements have been shown to help stem cells stay undifferentiated (a good thing) and to multiply.[11]

- Vitamin D$_3$, vitamin C, and green tea extract have a variety of positive effects on stem cell circulation, production, and response.[12]

- Tai chi exercises raised individual stem cell counts by *three to five times* in a study from China.[13]

- Getting high quality sleep helps keep stem cells young.[14]

STAY YOUNG DOWNSTAIRS WITH STEM CELLS AND OTHER TREATMENTS

The genital stem cell treatments that Lana and I both underwent may sound a bit extreme, but in reality, they were mildly uncomfortable for a short time and accomplished a lot more than just improving our sex life. To make it to a hundred and eighty without needing diapers, you have to maintain the integrity of the muscles, tissues, and blood flow in the genital region. Far too many people lose control of these muscles long before a hundred and eighty (or even half that age), and stem cell treatments are only one way I'm actively working to avoid this.

The nondrug strategies that help prevent or reverse erectile dysfunction (ED) do the same thing for incontinence. I have a lot of friends in their forties who are already popping blue pills, but many men see signs that things are headed in this direction even earlier. They're often too embarrassed to ask their doctors about it, but it's a lot easier to catch these signs and prevent ED and eventual incontinence than it is to reverse it. So let's talk about it.

If you're a man, the strength of your erection is a huge indicator of how well you are doing from an anti-aging perspective. As you read earlier, if your body is working well enough to reproduce, you are biologically young. Before you try to hack the problem directly, though, it's important to look at the underlying cause of ED or incontinence.

You can take a pill and get erections again, but that does not mean you've solved the core problem, and that core problem will continue aging you until you address it. For women, the signs of premature aging may be less obvious. Incontinence is primarily a pelvic floor issue in women, while sexual dysfunction is largely hormonal. In both sexes, a lack of sexual desire and/or function is a clear signal that something is off with your biology.

For both men and women, I recommend working with a functional medicine doctor to look at your hormone levels (particularly testosterone and thyroid hormones), blood sugar, blood pressure, and overall cardiovascular health. Issues with any of these systems can cause problems with sexual function.

Prescription medications are another major cause of sexual dysfunction, so talk to your doctor about whether or not any of your current medications may be creating or contributing to this issue. If so, work with your doctor to safely lower your dose or find an alternative. Don't just stop taking prescription meds cold turkey. As you read earlier, birth control pills are unfortunately a major cause of sexual dysfunction in women. If you are on the pill or another form of hormonal birth control, it's a good idea to talk to your doctor about alternatives that might be right for you.

Once you've addressed these issues, you might want to explore other ways of improving sexual function and/or incontinence. Besides stem cell treatments, one of my favorite technologies is GAINSWave, a treatment using shock-wave therapy that has been proved to effectively treat ED.[15] It uses high-frequency acoustical waves that break up the microplaque in existing blood vessels and stimulate the growth of new blood vessels and nerves when applied to the genitals. This leads to increased blood flow, which improves sexual function in both men and women.

I decided to try it because I'm interested in anything and everything that can improve performance . . . even in the bedroom. The results were astounding and are difficult to describe without getting a little too graphic, but the part about growing new blood vessels is absolutely real. My (ahem) equipment grew in length by more than 15 percent. It took me about three months to stop doing a double take when I glanced in the mirror while getting out of the shower. These "outsize" results may be because I had injected stem cells before I did the shock-wave treatment. In any case, wow!

GAINSWave is also useful in treating stress incontinence. As many women know, this can happen when physical movement or activity—such as coughing, sneezing, running, jumping, or heavy lifting—puts pressure on the bladder. This is common when pelvic floor muscles and other tissues that support the bladder and regulate the release of urine weaken with age or after childbirth. To become Super Human, you need to keep these tissues and muscles young whether you are a man or a woman.

If you aren't ready to invest in supersonic shock-wave or stem cell treatments, Kegel exercises are a simple alternative—for both women and men. To do Kegels, contract and release the pelvic floor muscles repeatedly for a few minutes each day. It's free and easy, and you can't even tell that I'm doing it right now as I type these words. Do it with me as you read! When you're a hundred and twenty years old, you'll be glad you did.

REGENERATE WITH BADASS CANCER-FIGHTING CELLS

In 2006, Japanese stem cell researchers achieved a major break-through in producing Wolverine-like healing powers when they cre-ated induced pluripotent stem cells (iPSCs) in a lab. These are adult stem cells that have been genetically reprogrammed to become "blank slates" with the potential to become any type of cell in your body, just like embryonic cells. In addition, researchers recently discovered that we already have very rare pluripotent cells in our blood that can divide and become any other type of cell. These are called very small embryonic-like cells (VSELs). They are typically inactive,[16] but they can be activated through culturing and ultrasonic vibration. One thing that makes VSELs so powerful is that they contain high levels of tissue inhibitors of metalloproteinases (TIMPs), compounds that increase neurogenesis, the birth of new brain cells.[17] You want that.

That's why in 2017 I had my own VSELs extracted, activated via mechanical vibration, and reinjected. This gave me more TIMPs, which probably gave me a better brain. VSELs are small enough to pass through the blood-brain barrier, but no one has proved yet that they actually go from your blood into the brain. Just in case they don't, I worked with Dr. Matthew Cook at BioReset Medical to go overseas, where we introduced my VSELs into my sinuses, very close to my brain, along with a little dose of nasal insulin to make them go in faster. Younger brain, here I come!

After the procedure, I noticed a shocking improvement in my visual acuity. I could see details in trees from very far away, and a vision test conducted a year later showed my vision was 20/15. (I do not have a recent vision test from before the VSELs to compare, but I don't believe my eyes were previously that Super Human.) Unless there is a convincing reason otherwise, I'll continue to do VSEL treatments once a year because they require only a blood draw and are quite affordable compared to stem cells—at least if you're already overseas or can find a cheap ticket.

I expect to see a lot more research come out in the next few years about using VSELs to treat degeneration and disease because they are cheaper and more versatile than stem cells. And we'll also learn a lot more from further study of induced pluripotent stem cells. For instance, researchers in California just figured out how to turn these "blank slate" induced pluripotent stem cells (iPSCs) into special immune cells called modified natural killer (NK) cells.

NK cells are the first-responder Special Forces of immune cells. They locate and identify infected or cancerous cells early and quickly respond by killing them. The thing that sets NK cells apart from other immune cells is that they can detect when cells are dysfunctional even before those infected cells trigger antibodies or inflammation. Other immune cells, such as T cells, are blind to dysfunctional cells until it may be too late—after inflammation has started. NK cells can therefore detect and kill tumorous cells before they turn into real tumors. You want more NK cells to keep cancer, one of the Four Killers, at bay. And now we can make them at will!

In animal studies, the modified NK cells displayed heightened activity against ovarian cancer.[18] The lead researcher conducting this study said that one batch of these special NK cells, which are created in a dish from mature human cells, can potentially treat thousands of people suffering from cancer. They are now working to manufacture iPSCs and begin human trials. There are also clinical trials going on right now to treat cancer patients using donated NK cells via transfusion.

Here's why this matters for you. If you live long enough, the odds are very high that you're going to get cancer. In fact, you probably have a few cancer cells floating around right now, but your immune system will likely take them out the way it's supposed to. But what if you could affordably get an IV every couple of years that would flood your body with NK cells to mop up any cancer cells your immune system may have missed? This is exactly what's happening now, which is great if your goal is to not die.

However, NK cells do a lot more than just hunt down cancer before it happens.

Very recently, scientists discovered a little-known component of cell membranes called perforin that helps get rid of old senescent (death-resistant) cells.[19] Of course the drug industry immediately began looking for a drug to increase perforin levels. Some are in clinical trials now. But it turns out that your NK cells actually produce perforin.[20] It follows that increasing your NK cells will cause perforin levels to increase and your number of senescent cells to decrease.

When I learned about this research, I decided it was a good strategy to extract and culture my own NK cells, something that is usually reserved for people suffering from certain types of cancer. I worked with Dr. Matt Cook again, this time to have my blood drawn during the Bulletproof Conference in 2017. He met me at a hotel in Pasadena and pulled vial after vial of my blood during an advisory meeting with a few slightly pale start-up entrepreneurs who pitched me on their crypto start-up as the vials filled. Dr. Cook shipped my blood off to an undisclosed lab, and six weeks later I learned that they had successfully cultured 2.07 billion NK cells—a huge amount. Then I traveled out of the country to have them injected via IV.

Did this increase my perforin levels and cause my body to dump senescent cells, shoring up one of the Seven Pillars of Aging? Almost certainly. Over the next ten days without changing my diet or exercise routines, I lost ten pounds, and my body fat dropped by more than

2 percent. What percentage of that weight came from senescent fat cells? The anti-aging doctors doing this type of treatment say it can help your immune system function like it is twenty years younger. Given my history of poor immune function, that's an important strategy for my longevity.

That said, this is an absurdly expensive procedure, and it's not even on the menu unless you know who to ask, though if you have certain medical conditions you can get infusions of donated or lab-grown NK cells. The reason I'm writing about it here is because the more people who know about these kinds of treatments, the more demand will grow and costs will go down. There's no reason this treatment should be so expensive and no reason you should have to leave the country to get it. Once this work is allowed to progress, costs will drop and it will be a standard anti-aging therapy that adds years to your life, or at least life to your years. Every single one of us could benefit from a better immune system, not to mention what it could do for people who are seriously ill.

Let's assume you're not going to grow your own NK cells in a lab for a couple more years. There are still things you can do to improve your existing NK function. For example, simply spending more time in nature, particularly surrounded by trees, can boost NK cells. Many evergreen trees give off aromatic compounds called phytoncides that increase NK cells. In fact, cypress essential oil in a diffuser at night significantly increases NK activity and perforin levels![21] Likewise, avoiding toxic mold will help, as mold breaks NK function.[22] Managing stress is also hugely important, as chronic exposure to stress hormones suppresses NK cell function. This is just one more reason that managing your stress is not optional if you want to become Super Human.

BATHE YOUR CELLS IN VIRGIN BLOOD

For hundreds of years, there have been stories of people who sought immortality via a slightly eccentric route: the blood of virgins. Take Countess Erzsébet Báthory de Ecsed, a Hungarian noblewoman who was born in 1560 and allegedly murdered hundreds of young girls.

Why? Bathing in the blood of virgins was an important part of her beauty routine, of course. And we all know that vampires had a taste for virgin blood, too.

Whether or not virginity has anything to do with healthy blood, it certainly seems that youth does. Doctors noticed many years ago that hemophiliacs, who routinely received blood transfusions, could tell the difference in how they felt after receiving blood from a young person as opposed to an old person. When they got blood from a young person, they felt more energetic. And while this is anecdotal, there's good science to back it up. A 2014 study in *Nature Medicine* reported that exposing an aged animal to young blood can counteract and reverse preexisting effects of brain aging at the molecular, structural, functional, and cognitive levels.[23] Sounds pretty good to me!

This led researchers to begin studying exactly which components of youthful blood are responsible for these effects. And it's led companies to begin buying blood from young people, such as college students, and then selling it to older folks, who receive it via IV. One could say this is morally ambiguous, especially since the treatment costs $8,000—of which the students themselves are getting a very small percentage. However, donating blood extends your life by decreasing your levels of ferritin, a blood cell protein found in iron. High ferritin levels are incredibly aging. So if college students are being paid and are also benefiting physically from their blood donations, I'm okay with it. (Not that it's up to me.) What I'm not okay with, however, is the fact that there are diseases carried in the blood that may not be detected by conventional testing. Additionally, that there are no actual studies to prove that "young blood" transfusions offer any real benefit, despite the fact that, at the time of writing, 104 people have paid a jaw-dropping $8,000 per transfusion.

Don't have eight grand to drop on young blood? That's okay. We already know about some of the substances that make it so rejuvenating. And it's much simpler and less expensive to supplement with those substances than it is to get a transfusion. I'm sure that within the next few years we'll discover much more about exactly what makes young blood so special. For now, we know about a few of its important components.

KLOTHO

One of the precious ingredients in young people's blood is Klotho, a protein produced primarily by the kidneys. It is named after Clotho, one of the three Greek goddesses who, according to mythology, determined how long each human being would live.

A Japanese physician named Dr. Makoto Kuro-o inadvertently discovered Klotho when he was conducting research in a different area and created mice that lacked the gene that told their bodies to produce Klotho. The mice lived only 20 percent of their normal life-span and died of apparent old age.[24] They had wrinkled skin and were very frail, with kidney failure and cognitive problems. After a few more years of experimentation, Kuro-o successfully created mice that produced more than the normal amount of Klotho and lived 20 to 30 percent *longer* than normal.[25] Don't you want that, too?

Recent research on elderly people shows that Klotho influences human life-span as well. (It's almost as influential as Clotho herself!) Studies show that seniors with the lowest Klotho levels have a 78 percent higher risk of death over a six-year period than those with the highest Klotho levels, even after correcting for sex, age, and health status.[26]

There are some naturally occurring variations in the Klotho gene. A quarter to a fifth of us have a single copy of a modification in the Klotho gene (the KL-VS variant), which results in higher levels of Klotho in the blood.[27] These individuals tend to live a long time. They also have larger than normal prefrontal cortexes and better cognitive skills than people with the standard Klotho gene.[28]

The kidneys produce the vast majority of the Klotho in the body, so patients suffering from kidney disease see a sharp decline in blood Klotho levels. But it's unclear whether the kidney disease causes low Klotho levels or low Klotho levels cause kidney disease. We do know that injecting synthetic Klotho into mice with acute and chronic kidney disease reduces kidney damage and fibrosis, slows progression of the disease, and promotes healing. Klotho injection also reduces damage to the heart and resulting heart failure, a common cause of death in patients with kidney disease.[29]

This has led scientists to investigate the relationship between Klotho and other diseases of aging. They've found that patients with Alzheimer's disease have decreased Klotho levels in their cerebrospinal fluid. And increasing Klotho levels in mice with Alzheimer's disease decreased brain cell dysfunction and ameliorated cognitive and behavioral deficits.[30] Klotho injection in both young and old mice also increased cognitive skills and learning ability,[31] while reducing tumor growth and metastasis in lung, breast, and prostate cancer.[32]

Further, patients with type 2 diabetes have decreased levels of Klotho, and a lack of Klotho is associated with decreased insulin production.[33] When diabetic mice were treated with Klotho for two weeks, they saw a significant decrease in blood glucose levels and an increase in insulin levels. Finally, older adults with lower than normal Klotho levels have poor skeletal muscle strength.[34] Researchers believe that Klotho suppresses aging in tissues throughout the body and stimulates muscle vitality.[35]

My friend Jim Plante, a longtime entrepreneur and former CEO of a genetic testing company, developed polycystic kidney disease and started a company called Klotho Therapeutics to synthesize Klotho and offer treatments. Of course I volunteered to be one of his first guinea pigs as soon as it is available, and I will likely have tried it before this book is in your hands. Until then, you can improve your own Klotho levels by:

- **Avoiding stress.** A high-stress environment substantially lowers Klotho levels.

- **Exercising.** This raises Klotho levels.[36]

- **Supplementing with vitamin D_3.** People who take this vitamin have increased Klotho levels.[37] (Always take D_3 with vitamin K_2 and vitamin A.)

- **Controlling your blood pressure.** Angiotensin II, the hormone responsible for increased blood pressure, also suppresses Klotho levels.[38]

- **Maintaining healthy hormone levels.** In particular, higher levels of testosterone are related to higher Klotho levels.[39]

COPPER PEPTIDES

Another one of the main components in young blood that makes it so powerful for anti-aging is GHK-Cu (copper peptide). This chain of amino acids is plentiful in your blood when you are young but normally declines as you age. Your body releases GHK after you get injured, which may help explain why younger people heal so much faster than older people . . . until now.

You can supplement with GHK that's not derived from child blood because it is easy to synthesize. Sadly, it is unlikely to be a focus of big-time research budgets because it can't be patented. You can buy and use GHK by applying it topically or injecting it into the muscle, via IV, or under the skin. This peptide attracts immune and skin cells to injury sites for rapid wound healing and increased collagen synthesis. In one study, GHK gel applied topically helped heal people's skin ulcers three times faster than a placebo.[40] GHK also reduces inflammation and scarring and acts as a powerful antioxidant.[41] Remember how chronic inflammation leads to tissue cross-linking? GHK to the rescue. You've probably seen skin-care ads mentioning copper peptide—it's the same stuff we're talking about here, just smeared on the skin instead of used internally.

GHK is also incredibly rejuvenating for your brain. It helps neurons grow faster and strengthens neural connections. In rodents, supplementing with GHK improves symptoms of dementia.[42] And because of its impact on collagen synthesis, GHK can tighten skin, improve elasticity, reduce fine lines and wrinkles, and improve discoloration from skin damage and aging.[43]

There's even some evidence that GHK injections can increase hair growth and help hair maintain its color,[44] two things I'm not ashamed to admit I think about when it comes to aging. To some extent, genetics influence both balding and grayness. My mom went completely gray in her twenties, and pretty much all the men in my family are bald. I started getting noticeably grayer within the last few years, and I wanted to stop this from progressing any further. I like to think it's not vanity—just a desire to look as young as I feel.

I started researching what actually causes grayness and found that one cause is a copper deficiency. So after I used a blood test to determine that I did not have copper overload, I developed a copper/zinc formula, and I have worked with physicians to inject GHK intravenously without any virgin blood. My number of gray hairs has reduced quite substantially, but you'll read all about that in the next chapter. GHK also caused some baby hairs to start growing back into my slightly receding hairline. I'm hopeful that by the time I'm a hundred and eighty I'll have a full head of hair that's not gray (without dyeing it).

I also use GHK topically with profound results. Look for a skin-care cream with at least 2 percent GHK for younger-looking skin or as a way to heal from cuts and bruises. My friend Andy Hnilo uses GHK copper peptides in his Alitura skin-care line, which helped him recover with almost no scarring from a horrific car accident that broke his jaw in five places. I use the Alitura Gold Serum every day because it contains GHK and other cofactors.

It's a good idea to take care of your skin, but you want to do it in combination with other techniques to age backward from the inside out. It sounds like a cliché, but beauty really does come from within. If your cells are biologically young, you will look young—end of story. So I strongly recommend pairing skin care with other anti-aging techniques to speed healing and age backward both inside and out, which leads us to the next chapter . . .

Bottom Line

Want to heal like a deity? Do these things right now:

• Spend more time in nature to boost your own natural killer cells and enhance your immune system; bonus points for frequently visiting a forest with lots of evergreen trees. Or at least use some forest-based essential oils like cypress.

• Consider boron supplements for stem cells, as well as the other listed stem cell enhancers. Calcium fructoborate or food-grade boron (tetraborate) work well.

• Make sure your sexual function is that of a young person. If it isn't, get your hormone levels checked and look at any prescription meds that may be causing a problem. To improve sexual function, consider GAINSWave treatments or simply practice Kegel exercises on a daily basis.

• Try using copper peptides topically or via injection or IV to enhance your body's ability to heal.

• If you want to pull out all the stops or have major injuries that need healing, look into available stem cell treatments. These aren't cheap, but they're often similar to the price of surgery without many of the risks.

DON'T LOOK LIKE AN ALIEN

Avoiding Baldness, Grays, and Wrinkles

Type the words "anti-aging" into any Internet search engine, and you can guarantee that the majority of results are links to skin-care products and ads for plastic surgery. As a culture, we seem to be more afraid of looking old than actually *being* old. Having experienced the metabolic state of an old person when I was young, I can say with authority that I'd rather look old if it meant I could feel young. But why not aim for both? The good news is that if you use the recommendations in this book to make yourself young inside, your outside will look younger, too.

In fact, you'll actually look your best when you stop aging at its source—your mitochondria—because the same cuts that age your mitochondria and cause inflammation also create the visible signs of aging. It's no big mystery—your skin and hair follicles are made of cells, which are powered by mitochondria. If those cells are full of waste products or can't produce energy efficiently, you are going to look old no matter how much money you spend on eye cream. And if you hack your skin and hair but keep eating inflammatory foods or exposing yourself to a toxic environment, you're not going to get the best possible return on your investment.

That's why this chapter appears toward the end of the book. There are specific techniques you can use to reverse visible signs of aging, but you'll get much better and faster results after you remove the rest of the hits from your environment. Remember the number one rule of biohacking—first remove the things that are making you weak (or old). Once you've done that, you'll get a much greater ROI on the techniques that follow.

BUILD NEW SKIN, JOINTS, BONES, AND GUT LINING WITH COLLAGEN

Maintaining youthful-looking skin as you get older means continuing to build youthful collagen. Collagen is the most abundant protein in your body, and it acts as a building block for your bones, teeth, muscles, skin, and all other connective tissues. Maintaining youthful collagen as you age isn't just about your skin. To stay young and regenerate like a badass you need healthy collagen in all of your muscles and tissues, too.

Your body contains at least twenty-eight different types of collagen. But the vast majority (80 to 90 percent) of the collagen in your body is type 1, type 2, or type 3. Types 1 and 3 provide structure for your skin, muscles, and ligaments, while type 2 is found in cartilage and your eyes. Collagen makes up nearly 80 percent of your skin, and it's found in the middle layer called the dermis. It works together with another protein called elastin to strengthen the skin and help it snap back into place when stretched. Sagging, crepey skin is a clear sign of a lack of healthy collagen.

As is true of most of the interventions in this book, the younger you are when you start working on building healthy collagen, the better. Taking care of your skin when you're young is a much more effective way to avoid aging than trying to reverse the damage later. We know that the half-life of collagen in skin is fifteen years,[1] so if you start supplementing with collagen itself or collagen-building supplements now, in fifteen years, half of your collagen will look a lot better than it would have otherwise.

This is a must because collagen production slows down as you get older—and it happens sooner than you'd think. After the age of twenty-five, you break down more collagen than you build, and this is often when you begin to see your first fine lines and wrinkles. From there, you start losing about 1 percent of your collagen each year.[2] And that's just the average. Excess sun exposure, smoking, and too much UV light exposure all degrade collagen even more quickly.[3]

One percent might not seem like a lot, but if you join me on the quest to live a long time and don't hack those statistics, you'll end up with exactly 16.38 percent of your collagen left on your hundred and eightieth birthday. While it might be kind of cool to be able to look through your skin at your own liver, there are better things you can be doing at that age. If you slow your rate of collagen loss by 50 percent, you'll have 2.5 times more collagen in your skin when you're a hundred and eighty. And if you occasionally stimulate new collagen to grow, you're going to like how you look as each year passes.

This is one reason I worked so hard to bring collagen protein powder to the market and popularize it when it was unknown. Today it's all the rage, and for good reason. Collagen protein powder is hydrolyzed, meaning it's been partially broken down into the main amino acids that your body needs to produce more collagen: glycine, proline, and hydroxyproline, along with some smaller fragments called peptides. Supplementing with these can help your skin look younger, for sure. Research shows that collagen supplements improve skin elasticity, reduce wrinkles, boost skin hydration, and increase the density of fibroblasts, the cells in connective tissue that produce protein.[4]

Collagen supplements do a lot of other things, too. They can reduce joint pain and boost the density of your cartilage, making joints more flexible. A 2008 study found that athletes who took hydrolyzed collagen for six months saw a significant improvement in joint pain.[5] Another study showed that men and women over fifty who took collagen for six months experienced a decrease in spinal pain.[6]

Taking collagen supplements is an easy way to help prevent age-related arthritis, and it can help you avoid osteoporosis, the thinning and weakening of bones with age. Postmenopausal women are especially vulnerable to osteoporosis because of a lack of bone-protecting estrogen. A 2018 study of these women showed that taking collagen supplements for twelve months increased the amount of minerals—namely calcium—in their bones, making them stronger.[7]

Collagen also repairs and strengthens the intestinal and stomach lining.[8] A stronger lining can help reverse leaky gut syndrome, which you read earlier is incredibly aging, and can make it easier to absorb important nutrients. The amino acid glycine that makes up a third of collagen also helps your body produce more stomach acid, which aids in digestion and reduces acid reflux.[9]

Quick sidetrack about collagen and acid: When I was in my twenties I had terrible heartburn, something you might think is reserved for older people. I went to the doctor and said, "I feel like I have a candle burning in my chest." He gave me an acid blocker, which worked temporarily. But it turns out that a lack of stomach acid actually *causes* gastric reflux by sending a signal to your body to not close the sphincter at the top of the esophagus. Acid can then flow into the esophagus, causing the pain associated with heartburn. When you have enough stomach acid, the sphincter closes. Your body needs stomach acid for other reasons, too, like to sterilize the food you eat and break down its proteins and fats. Acid blockers relieve symptoms temporarily, but as soon as you stop taking them, the pain comes back stronger than ever, and they wreck your ability to absorb nutrients.

There is a natural substance called betaine hydrochloride (HCL), which can supplement the body's natural hydrochloric acid (stomach acid) without negative effects. I stopped taking the acid blockers and began taking betaine HCL with meals. The best way to use this supplement is to find the number of capsules that actually make heartburn worse, and then back off from that number by one. Take it at the beginning or middle of a meal, not at the end.

Your levels of stomach acid decline as you age, from about 180 mg at twenty years old to about 50 mg at over sixty.[10] One study found that 30 percent of people over the age of sixty have almost no stomach acid secretion, and another found that 40 percent of postmenopausal women have none.[11] In my twenties, I had the stomach acid secretion of someone three times my age.[12]

What does this have to do with collagen? Remember, glycine is one of the main amino acids in collagen and it helps your body produce stomach acid. Perhaps as a young man I was deficient in collagen and therefore glycine, but more likely I just ate too much sugar.

Glycine is actually an inhibitory neurotransmitter, which means it calms the nervous system and can help you get better quality sleep. One study found that when people who normally had trouble sleeping took glycine before bed, they fell asleep quicker, slept more deeply, and experienced less daytime drowsiness.[13] That study led me to write the first blog post ever about using collagen before bed as a sleep hack, and today you'll find it echoed across the Internet.

In addition to the necessary amino acids, including glycine, your body needs sufficient vitamin C to produce collagen and maintain its vitality. Vitamin C works in two ways to support your skin. First, it is a powerful antioxidant, so it protects your skin cells from damaging free radicals that break down collagen. Second, you need vitamin C to make and repair collagen. Having enough vitamin C ensures that your body can assemble amino acids into collagen when you need to. You can get skin benefits by eating vitamin C–rich foods, taking a vitamin C supplement, and/or applying a vitamin C serum topically.

Once you have the raw materials you need (the right amino acids and vitamin C), there are several specific things you can do to help your body produce more collagen.

CRYOTHERAPY

Brief cold exposure, or cryotherapy, increases collagen production and blocks the inflammatory enzymes and hormones that destroy the collagen you already have.[14] Standing in air that is cooled to 270 degrees below 0 F for up to three minutes sounds worse than it is. Steep temperature drops like this increase your production of the antioxidants glutathione and superoxide dismutase, which help fight the free radicals that age your skin.[15] It also causes your body to burn extra calories. You can also get cryofacials, which are basically cryotherapy for your face. A small stream of cold nitrogen gas is directed at your face to increase blood flow to the skin. This boosts collagen production, as your blood carries the raw materials your skin cells need to produce collagen.

This all might sound shockingly uncomfortable, but the more accessible alternative is *far* more uncomfortable: the humble cold shower. Cold air is not such a big deal, and on your face it actually feels refreshing. Cold water feels a lot worse, but it is effective. Cold exposure helps mitochondria, and it also stimulates capillary beds to increase circulation in your skin. Just aim the coldest water that will come out of your shower at your face and neck area for one minute. It will be unbearable for exactly three days, and that's if you can even do it for a full minute. After that, your mitochondria change their amount of cardiolipin, a component of the inner mitochondrial membrane, so they can produce heat more quickly. The side effect is they can also make energy better, and suddenly that cold shower goes from painful to strangely relaxing and invigorating. Your collagen will thank you if you decide to experience three days of discomfort.

MICRONEEDLING

You can buy a microneedling roller for less than $20 that uses tiny needles to prick invisible holes in the top layer of your skin. This disrupts collagen and stimulates the body to form new, youthful collagen fibers. It's basically a form of hormetic stress for the face. Your skin cells realize, "Dang, I guess there are going to be needles sometimes," so they toughen up and get younger and stronger while weak or damaged skin cells are killed off. The results are pretty powerful. One study found that nearly 100 percent of patients with deep facial scars showed significant improvement after just three microneedling sessions.[16]

To take this a step further, you can combine microneedling with platelet-rich plasma (PRP) in a treatment known as the vampire facial. To do this, a doctor takes your own blood and separates out the plasma, which is rich in growth factors. The plasma is applied to your face before and after microneedling to further stimulate the growth factors triggered by the punctures. Aestheticians usually do this with a clinical-grade electric microneedle and numbing cream. It may sound crazy, but it works. A 2014 study found that the vampire treatment improved acne scars more than just microneedling alone.[17]

You can get your own electric microneedle for about $100, which works better than a $20 roller. If you try either one, make sure everything is sterilized and free of contaminants every time you use it. Whenever you puncture the skin, there is obviously a small risk of infection. These rollers work on your hairline as well, and can stimulate hair growth. They are cheap, they work well, and they don't take long to use. Totally worth it.

RETINOL

In addition to copper peptides (which you read about in the previous chapter), retinol is one of the most effective ingredients for your skin. Retinoid is the blanket term for a pure form of vitamin A. Some formulations like Retin-A require a prescription because they contain high amounts of retinoic acid, the active ingredient in retinoids. This acid purges old skin cells and causes new, healthy cells to form quickly.[18]

Over-the-counter products known as retinols are less potent because the body needs to convert them into retinoic acid. This extra step means retinols take longer to work compared to prescription products, but they are still effective. Retinol causes skin cells to turn over quickly and increases collagen production,[19] leading to a reduction in fine lines and wrinkles, faded age spots, smoother skin, increased firmness and elasticity, and smaller pores.

Sounds pretty good, but there are a few downsides. If you have sensitive skin, you may find retinol irritating. You should always wear sunscreen when using retinol, since it makes your skin more sensitive to the sun. And pregnant and breastfeeding women should avoid retinol, since in high doses it can harm a fetus's or a baby's development.[20]

If you are pregnant, nursing, sensitive to retinol, or simply want a gentler alternative, there is one plant-based ingredient—bakuchiol—that has been proved to mimic the activity and benefits of retinol. Bakuchiol comes from the seeds and leaves of the *Psoralea corylifolia* plant and is traditionally used in Ayurvedic and Chinese medicine to treat skin diseases. In one study, people who used bakuchiol twice a day for twelve weeks saw a significant improvement in fine lines and wrinkles, pigmentation, skin elasticity, firmness, and collagen production.[21] Best of all, bakuchiol offers all of these benefits minus the dryness and flakiness typically produced by the use of retinol.

METHYLENE BLUE

If you read *Head Strong*, you're already aware of the fact that methylene blue, a medication and blue dye, is a powerful cognitive enhancer. It acts as an antioxidant to mitochondria, meaning it protects them from aging oxidative stress. And because it is absorbed through the skin, it can also protect connective tissue cells from aging. A 2017 study showed that methylene blue was more effective at delaying skin cellular senescence than other antioxidants. The same study demonstrated that methylene blue improved skin viability, promoted wound healing, and increased skin hydration and dermis thickness while upregulating production of elastin and collagen.[22] Not bad for a very affordable chemical that's been around for decades!

The problem is that very few companies make methylene blue skin-care products, and it can't be patented. It is far easier and more affordable to buy a dropper bottle of medical-grade or food-grade methylene blue (skip the chemical supply or aquarium cleaner versions) and put a few drops into the other products you put on your skin. If you put too much, you'll look like a Smurf. If this happens, you can usually wash it off with soap as long as you catch it fast enough. I add a few drops of methylene blue to my copper peptide serum and body lotion, and it makes a noticeable difference. This is an incredibly powerful skin booster that is painless and inexpensive. And since you just use a few drops at a time mixed into your other skin-care products, a $20 bottle will last you for a couple of years. I saw the difference after using it on my face for less than a week.

LASER FACIALS

We've already discussed some of the skin-enhancing benefits of light therapy, including red light and near infrared light. There are also common and effective procedures using lasers to resurface the skin and make it appear younger. Ablative lasers remove tiny layers of skin to reveal a youthful glow, and less invasive non-ablative lasers stimulate collagen growth and tighten the skin. Before trying an ablative laser treatment, be aware that you may experience pain and swelling after, and it can take several weeks for the skin to fully heal. Don't try this on the day before an important event! But the results are powerful and can last for years. Non-ablative facials require less recovery time (though your skin may be red and swollen for a short period) and are effective, though the results are not as impressive and do not last as long as the more invasive lasers.

MELANIN

Your brain produces a little-known hormone called alpha-melanocyte-stimulating hormone (alpha-MSH). Its job is to send a signal to your melanocytes, the cells that make the pigment melanin that gives skin and hair their color and protects cells from damage that can lead to aging and even skin cancer.

It turns out that alpha-MSH is a broad-spectrum anti-inflammatory hormone, and people with autoimmune conditions and those who have been exposed to toxic mold (like me!) tend to have lower than normal levels.[23] After my lab tests confirmed I was low in alpha-MSH, I bought myself some and injected a little once or twice a week. This is not without risk. There is some evidence that very high doses may increase the likelihood of melanoma, but there is also evidence that it can help prevent cancer. Given the low-dose, occasional use, and the other things I do to slash my cancer risk, I decided to use it.

Aside from making my skin look better and giving me a great tan without a lot of sun, the alpha-MSH should increase the melanin levels in my eyes and brain. Our eyes and brains require melanin to efficiently produce energy. As you read earlier, when exposed to sunlight or mechanical vibration, melanin has the power to break water apart, freeing up oxygen and electrons for your mitochondria to use to produce ATP (energy).[24] One major side effect from injecting alpha-MSH is that thanks to all this extra energy it makes Viagra look weak. If you're a man and you use it, expect to feel like a teenager the next day.

You can also produce more melanin by consuming extra polyphenols. Your body makes melanin by linking polyphenols together from vegetables, coffee, tea, and chocolate. You read earlier that polyphenols are hugely beneficial for your gut bacteria, so eating more of these foods is a win-win for your longevity inside and out.

THE LONGEVITY OF YOUR HAIR

Melanin also plays an important role in making sure your hair continues looking young as you get older. Melanocytes (pigment cells) in your hair follicles produce the melanin that gives your hair its color. As melanin production declines with age, grays start to pop up and slowly take over.

In 2009, a team of European scientists made a big breakthrough in determining what exactly causes our follicles to produce less melanin. They found that hair follicles produce a tiny amount of hydrogen peroxide, which builds up over time and damages melanocytes.[25] Why does it build up? When you are young, a hardworking enzyme called catalase breaks down hydrogen peroxide into water and oxygen. But as you get older, catalase production starts to slow down, and hydrogen peroxide starts to accumulate in the body. While it's nice to believe that gray hair is a sign of wisdom, it is really a reflection of damaged melanocytes caused by a lack of catalase.

Catalase is one of the body's most potent antioxidants, and other antioxidants can also help break down hydrogen peroxide. For instance, glutathione, the body's master antioxidant, breaks hydrogen peroxide down into water.[26] This is yet another reason to supplement with glutathione. It's also a good idea to eat more catalase-rich foods such as broccoli, cucumbers, radishes, and celery.

You can also ramp up your catalase production by taking antioxidants like ashwagandha, curcumin, saw palmetto, and vitamin E. A 2017 study showed that ashwagandha, an Ayurvedic herb, protected white blood cells in rats from free radical damage caused by hydrogen peroxide.[27] And there is real evidence that this can help prevent grays in humans. A scientific review of ashwagandha showed that when middle-aged men took 3 grams in powdered form every day for a year, they significantly increased their levels of hair melanin.[28]

Very recently, researchers at the University of Alabama at Birmingham have found a connection between gray hairs and viral infections.[29] In mice, they noted that when a stressor such as an infection triggered the immune system, it responded by attacking a gene called MITF, which helps melanocytes function properly. A shortage of

MITF triggered the immune system to further attack the melano-cytes, which led to gray hair. Until we can turn that gene on or off, your best bet is to maintain a healthy immune system to fight off viruses and do everything you can to avoid autoimmunity. Some of the more advanced anti-aging experts I know believe viruses are causing far more problems than we believe and take an antiviral drug like acyclovir every year. It has side effects, though, and the jury is still out on whether the risk-reward ratio is a good one.

When it comes to aging and hair, though, I'm not ashamed to say that my biggest concern is balding. I'd love to still be rocking a full head of hair when I'm well over a hundred, so I've done quite a bit of research on how to prevent and reverse hair loss. Many people think that hair loss with age is strictly a men's issue, but that is not the case. A huge number of women experience hair loss as they get older. And in both men and women, this actually begins when you're still young. Eighteen percent of men under the age of twenty-nine have moderate to advanced hair loss, and that number jumps to 53 percent for men in their forties.[30] Meanwhile, between 15 and 20 percent of women under the age of fifty experience hair loss, and the numbers go up from there.[31]

Like so many symptoms of aging, hair loss boils down to your hormones and your mitochondria. Many different hormones play a role in hair growth. Getting the balance perfect with the help of a functional medicine or anti-aging doctor should help you avoid hair loss and even regrow some new hair. It makes sense, then, that the only FDA approved drugs to fight hair loss actually do so by affecting your hormones. Rogaine, Minoxidil, Propecia, and finasteride all work by blocking an enzyme that is responsible for converting testosterone to dihydrotestosterone (DHT). Too much DHT causes hair follicles to shrink, which eventually leads to baldness. In addition to pharmaceutical drugs, there are several DHT-blocking shampoos on the market. You're better off with the shampoo, because oral drugs for hair loss often have unfortunate side effects, like shutting down *all* hormone function.

However, targeting DHT is only one way to fight off baldness. The hair follicle is a very sensitive mini-organ that requires a tremendous amount of energy from your mitochondria. You can have all the moving parts and raw materials you need, but you're never going to manufacture hair if you don't have the power behind that machinery to get the engines running.

In 2018 researchers at the University of Alabama at Birmingham turned on a mutation that caused mitochondrial dysfunction by adding the antibiotic doxycycline to the food and drinking water of mice. Just eight weeks later, the previously healthy mice developed gray and significantly thinning hair and wrinkled skin. But once the researchers stopped feeding doxycycline to the animals and their mitochondria got back to doing what they do best, the mice regained their healthy, youthful appearance within just four weeks.[32]

To avoid baldness, you need to look at the factors that may be causing your mitochondria to become sluggish. You already know what these are: stress, toxins accumulating in the liver, hormonal imbalances, inflammation, and of course free radicals. All of those factors downregulate the two major hormones that modulate mitochondrial activity—T3, which is a thyroid hormone, and progesterone.

First and foremost, stress messes with your thyroid hormone levels, which is a disaster for your mitochondria. When you are stressed, you begin overproducing the stress hormone cortisol. This inhibits thyroid-stimulating hormone (TSH), which in turn inhibits the production of another thyroid hormone called thyroxine, or T4. To use T4, the body must first convert it into the active hormone triiodothyronine, or T3, also known as the "energy hormone."

If your body cannot effectively convert T4 to T3 or you don't have enough T4, you convert it instead to an inactive hormone called reverse T3, or RT3. An improper T3/RT3 balance can effectively shut down the body by preventing you from producing enough energy. Many patients with symptoms of hypothyroidism appear to have normal levels of thyroid hormones. But this is just because most conventional doctors test only for TSH and maybe T4. If you have symptoms of low thyroid, including hair loss, insist on a T3/RT3 test as well. This all means the idea that stress can make your hair fall out is not an old wives' tale. It can happen when stress causes your body to make more RT3 and less T3 and your mitochondria can't produce enough energy.

Thyroid hormone disruption also causes hair loss, since these hormones play a major role in activating stem cells. There's a reservoir of stem cells in the hair follicle bulge. When these stem cells get a signal from thyroid hormones, they become activated and grow into new hair follicles. If stress is messing with your thyroid hormones, this signal is never sent, compromising new hair follicle production.

Researchers have begun looking at this signal from thyroid hormones as a possible opportunity for intervention. This signaling pathway is known as Wnt. When researchers injected growth factors into the hair follicles of twenty-six men, the Wnt pathway turned on, allowing the signal from thyroid hormones to stimulate the production of new hair follicles. The men saw an increase in hair shaft thickness and density after just one injection.[33] Since you can't buy this injection (yet), the next best thing is to use a Chinese herb called Danshen that upregulates Wnt[34] and helps with cardiovascular issues. So you can get your hair back and reduce your risk of one of the Four Killers.

You read earlier that I've been on thyroid medication since my mid-twenties. Two years ago, I attempted to go off of it. I maintained my energy thanks to all my anti-aging work, but developed a bald spot for the first time. I went back on the thyroid medication, used the interventions above, and my hair filled almost entirely back in! There is a reasonable argument that anyone over age fifty without documented *high* thyroid levels could benefit from trying a very low dose ($^1/_8$ to $^1/_4$ grain) of glandular thyroid medication containing both T3 and T4. Even slightly lower levels of thyroid hormones can contribute to increased fatigue and irritability and may also make it hard to lose weight. As you age, it is common to experience diminished thyroid hormone production.[35] And while having enough thyroid hormones will help your hair, as you read in chapter 9, the bigger story here is that a slight thyroid dysfunction predicts your risk of heart disease and death.[36]

Further, high levels of RT3 and not enough T3 leads to downregulation of progesterone, which causes estrogen dominance, an aging metabolic state in which estrogen levels far outweigh progesterone levels in the body. Many hair-care products compound this problem because they contain chemicals such as phthalates, parabens, and benzophenones that mimic estrogen in the body and throw off your hormone balance even more.[37] Estrogen plays a central role in regulating collagen production. With disrupted estrogen levels, you don't make collagen efficiently, and your hair, skin, teeth, nails, and joints all suffer.

Hormonal birth control products that contain high levels of estrogen also decrease the effectiveness of progesterone. This leads to a greater conversion of testosterone to DHT and can contribute to hair loss. When pharmaceutical companies discovered this, they began adding synthetic progesterone to birth control pills to counter the side effects, but synthetic progesterone is not the same as bioidentical progesterone. The synthetic version can increase levels of sex hormone binding globulin (SHBG), which you read earlier decreases your available thyroid hormones. This brings us right back to where we started: with too little thyroid hormone and thinning hair.

Besides stress, mitochondrial damage, and imbalanced hormones, the main culprits behind hair loss are environmental toxins, which compromise liver function. This matters because it's the liver that produces the T3 you so badly need for healthy hair growth. Glutathione, the master antioxidant, can help detoxify the liver, but first you must remove heavy metals from the body.

To maintain healthy hair, it's obviously essential to fight off hair loss at the root. (See what I did there?) And it's critical to make sure you aren't using toxic personal care products that will age you and your hair more quickly. I highly recommend doing a sweep of your medicine cabinet for any products containing phthalates, parabens, and benzophenones. It will also help to clean up your diet so you avoid the hormones in conventionally raised meat and the hormone-disrupting pesticides in conventionally grown produce.

If none of that works, how about a simple massage? Getting more blood flow to your scalp keeps hair follicles alive. There are some nice handheld rechargeable head massagers available today. They cost about $40, feel incredible, and will help to keep your scalp and hair in tip-top shape. Look for the ones with four little heads that move independently. The kind that looks like a wire whisk will not have the same effect.

At the end of the day, since mitochondria drive production of new collagen and lustrous hair, anything you do to improve your mitochondrial function will also help you look younger. This is a win-win that will hopefully keep you looking and feeling like a true Super Human for as long as you desire.

Bottom Line

Want to heal like a deity? Do these things right now:

For Skin

• Supplement with grass-fed or pastured collagen protein—at least 10 grams per day. It's available in unflavored protein powder, smoothie mix, ready-to-drink collagen Bulletproof Coffee, and collagen protein bars. You can also make bone broth if you don't like collagen protein.

• Eat more foods containing polyphenols and antioxidants: vegetables, coffee, tea, and chocolate. You can get skin benefits from vitamin C by eating vitamin C–rich foods, taking a vitamin C supplement, and/or applying a vitamin C serum topically.

• There is good science behind the skin benefits of cryotherapy, microneedling, and products containing retinol, copper peptides, and methylene blue.

• As you read earlier, red and yellow light therapy both have profound skin and hair benefits. See chapter 5 for a refresher. If you have significant skin damage or scarring, look into laser resurfacing.

For Hair

• Stop using chemical-laden personal care products and switch to all-natural versions. Throw out anything containing phthalates, parabens, and benzophenones. And if you are a woman, consider alternatives to hormonal birth control.

• To avoid grays, ramp up your catalase production by taking antioxidants like ashwagandha, curcumin, saw palmetto, and vitamin E.

• For baldness, try a DHT-blocking shampoo instead of prescription meds that have unwanted side effects.

• Deal with your stress, already! Seriously. If the threat of the Four Killers wasn't enough, maybe avoiding baldness will finally motivate you. This is not optional.

- If you are balding prematurely, get your thyroid levels tested by a knowledgeable anti-aging doctor, and make sure to check your levels of T3/RT3.
- To stimulate blood flow to the scalp, get a head massage or purchase an at-home massager.

HACK YOUR LONGEVITY LIKE A RUSSIAN

The use of performance-enhancing drugs is a touchy subject. Many people consider it cheating, and the most famous varieties of these drugs—like steroids—can be seriously dangerous. But there are some classes of these drugs that can do a lot for your longevity as well as for your performance right now. I have done the research and carefully considered exactly which substances I want to put into my body. And I'll let you know when I'm a hundred and eighty if it was worth it. So far I'm still alive and have been able to enjoy some pretty remarkable benefits without suffering from any major negative side effects.

PEPTIDES

Most, but not all, of these new substances come from peptides, which are compounds consisting of two or more amino acids linked in a chain. It's easiest to think of amino acids as letters. When grouped together in the right formation, they make words. These words are peptides. And like words, your body uses them to communicate with itself. When you combine several peptides, you have a polypeptide, which is a part of a protein and functions like a sentence. And a whole protein, like collagen, is essentially an entire paragraph. It consists of multiple letters strung together in just the right order to convey a clear message. In 1972, the year I was born, scientists discovered some of the many peptides that communicate messages to the body. For example, some tell the body to build muscle. Others say to enhance glucose metabolism or mitochondrial function.

Earlier you read about Epitalon, a peptide that strengthens telomeres, and GHK, or copper peptides. These are powerful substances. There are also bioregulator peptides that are targeted to keep specific organ systems young, and those with incredible healing powers. Unfortunately, we don't have the same number of double-blind placebo studies backing these compounds as we do for most drugs. We probably never will, but there is sound science behind all of it, and it's been in use for more than a decade. You might want to wait for more studies. But if you're going to negotiate with death, you have to weigh the evidence and risk against the potential rewards. We already know what will happen if you don't try anything new: You'll get old and die, most often after spending your last decade or two in painful decline. I'm willing to gamble, with admittedly high stakes, if it means potentially cheating death, even for a little while, and maybe becoming Super Human along the way.

BIOREGULATOR PEPTIDES

A great deal of research into bioregulator peptides has come from the gerontology labs of Vladimir Khavinson, a member of the Russian Academy of Sciences. Since 1987, he has been discovering which bioregulator peptides in young animals keep different organ systems young. He specifically focuses on how to prevent the loss of protein in critical organ systems. Most of these bioregulator peptides are extracted from animals and are called cytomaxes, but there's nothing stopping scientists from synthesizing similar peptides from scratch.

Synthesized bioregulators are called cytogens. They usually work quickly and are quite affordable—about $60 for a month's supply. You take them only every few months. The most interesting bioregulator peptides are the ones that can help to stop the Four Killers. Crystagen is designed to restore immune function, which can help with all Four Killers; Vesugen is meant to improve blood vessel protein synthesis; Pinealon is for better protein synthesis in the brain. There are dozens of others.

Being a professional biohacker, in 2017 I bought every single bio-regulator peptide there is and used them at high doses for sixty days. Some effects were readily apparent—the impact of the male hormone formula was very noticeable—but it's harder to know for sure if your vascular system is synthesizing protein better. Given the strength of the research on these and the relatively low cost, I'll be doing one course of bioregulators per year. I would inject them if injectables were available, but right now they come as capsules.

HEALING PEPTIDES

I've been experimenting with these for the past six years—here are the ones that have given me the best results.

TB500

Also known as thymosin beta-4, this peptide was quite effective for me. It is the peptide made by the thymus gland, which produces T cells and is an important part of your immune response. As you age and gain more senescent T cells, your thymus gland function normally decreases. This is one big reason your immune response usually weakens with age.

I've known that my thymus gland was trashed since the nineties. Since then, many of the functional medicine doctors I've visited recommended taking thymic protein, which is meant to boost thymus gland function, but I never felt a difference when I took it. TB500 was a different story. Research shows that it promotes wound healing and decrease inflammation,[1] helps you regenerate blood vessels,[2] and increases stamina. Following what you've learned so far, reducing inflammation helps you age better, and regenerating blood vessels is important to avoid cardiovascular disease.

The Russians have known that this stuff prolongs human life since 2003, when they studied the effects on 266 older people in a trial that lasted six to eight years. The researchers came to the conclusion that these peptides are promising anti-aging medicines that effectively treat diabetes, gastritis, and gastric ulcers; prevent cancer; treat infertility in men and women; and normalize immune system function.[3] Not bad.

In 2013 I ordered vials of TB500 online, and it arrived in a powdered form. Knowing that it works best via injection, I got a vial of sterile water, pulled some water out, and injected it into the powder, transforming it into a liquid. Then I wet my arm with an alcohol wipe and used an insulin syringe to inject myself with 5 mg. I did this once or twice a week.

This was not the first time I injected myself. Back in the late nineties, when I found out I had an issue with vitamin B_{12}, I couldn't afford to go to a doctor for the injections. So I bought needles from eBay and a tray of vitamin B_{12} injectables from a body building site. I had no idea what I was doing. This was before we had YouTube videos or I was married to a doctor. I did as much research as I could, washed my hands, and got to work. I cracked open the glass ampoule and used an inch-long needle to draw up a bright red fluid. Then I pulled my pants down, figuring I would inject it into my thigh muscle.

I knew what I needed to do, but I still sat there for almost an hour just staring at the needle. Every time I went to move the needle, my hand just stopped. It was nearly impossible to override my body's instinctive desire to not have that needle penetrate my skin! It took a ridiculously long time, but I finally did it. To my amazement, the needle slid in without pain or resistance. And once I had done it the first time, I never had a problem injecting myself again.

In situations like this, the body sends a message to your brain that you're going to die. It's very difficult to ignore that message, even though your brain knows better. The very act of overriding that fear response and choosing the brain over the body is incredibly empowering.

Years later, I spoke at a Tony Robbins event where there was a hot coal walk. The very point of this exercise is to override the body's fear response and walk over the hot coals your body fears but that you know on a cognitive level will not kill you. After injecting myself so many times and using neurofeedback to pare down my fear response even more, I was able to walk right up to the coals and keep going.

By the way, your body is not entirely wrong. Needles can be dangerous, and messing around with injections can actually kill you if you make a mistake. It's better to find a doctor who will do it for you.

BPC157—HEAL LIKE WOLVERINE

No, in this rare instance, BPC does not stand for Bulletproof Coffee! This is another powerful peptide that reduces inflammation and promotes rapid healing.

Ten years ago, I had a four-day-a-week yoga practice that I since let slide after having kids and doubling down on Bulletproof. I recently went back, and at my first class I kicked back from an arm balance called crow pose into plank pose, which is like the top of a push-up position. This is an advanced trick that I hadn't done in years, and I was pleased that I could still do it. When my feet landed, though, I jammed my toe hard enough that I chipped a small fragment of bone. This relatively minor injury continued to hurt for months because there was a loose fragment of bone in my toe that kept reinjuring the surrounding tissue.

I ordered the peptide BPC157 and injected it into my toe. Studies show that it promotes tendon and ligament healing[4] and even helps rabbits heal segmented bone defects.[5] It also heals the gut lining[6] and repairs damage stemming from irritable bowel disease.[7] The toe recovered quickly after a few injections and some dry needling to break up scarring. Tons of athletes and biohackers are using this stuff to help nagging injuries heal, and it's far cheaper than stem cell treatments.

Aside from helping you heal injuries like Wolverine, BPC157 can do wonderful things for your gut health. For gastric healing, just mix it into sterile water as if you were going to inject it, but place the same dose under your tongue and swallow it, instead. Reports of Crohn's disease and IBS quietly going away after doing this are not rare.

Some physicians recommend between 100 and 250 mcg once or twice per day, orally or via injection.

SARMS

The last ten years have given rise to an exciting new class of compounds called selective androgen receptor modulators (SARMs). The limited research on SARMs looks promising. They appear to build muscle and burn fat at a level comparable to steroids, but without the ball-shrinking, rage-inducing, liver-destroying, unsightly body-hair-growing effects. SARMs act on your hormones, but in a very targeted way, and they can help you rapidly build muscle and shed fat.

Like with most peptides, many of the studies that have so far been conducted on SARMs have used rats, and there haven't yet been any long-term human experiments looking at the safety of SARMs. There could be side effects we don't know about, and I acknowledge that playing with your hormones is risky. At the very least, though, SARMs are interesting compounds that merit discussion and consideration if you really want to go all in on anti-aging.

Keep in mind that SARMs are on the World Anti-Doping Agency's list of banned substances for athletic competition. If you're a competitive athlete, you shouldn't take these. Of course, the fact that they've been banned by most global sporting organizations means they actually work. Is this cheating? It's not for me to say. We live in a morally ambiguous world. On the one hand, we want athletes to perform at their very best possible levels. To do this, we're willing to let them wear special aerodynamic clothing and undergo bizarre training regimens that cost hundreds of thousands of dollars. On the other hand, if these athletes want to gain control of their biology by increasing levels of a peptide that will help them recover more quickly, we consider it cheating and make it illegal.

Personally, I think withholding an intervention that can help anyone live better is cruel. Years of competing take a toll on an athlete's biology. I have spoken with at least a dozen of the biggest names in sports—living legends who are suffering from the damage they did to their bodies while competing. Athletes know they are either one injury or a few years of aging away from involuntarily ending their careers. We have the technology available to keep them young, healthy, recovering, and even competing. They *want* to use it. But if they do, they will be punished.

In my opinion, there is no moral or ethical reason to ban these substances. We tell ourselves these rules protect athletes who might harm themselves by taking a risky drug. But we know that excessive exercise will also shorten your life, not to mention the long-term damage that accrues from recurring head injuries or crashing into a barrier at high speeds in a race. So why not let these people take something that will actually help them recover? It should be a basic human right to do whatever you want with your biology as long as you're up front about it and do it under the care of a knowledgeable physician. Then we can all learn from the world's best.

Bottom line: If you're not a professional athlete but rather a curious self-experimenter looking to upgrade your physical performance, SARMs may be worth considering. I repeat—these are not the same as synthetic steroids. Using underground steroids to boost your hormones is like trying to tweak a microchip with a sledgehammer. Synthetic steroids put on muscle, which makes them anabolic. Unfortunately, synthesized anabolic steroids negatively interact with your liver, your prostate, your heart, your sex organs (which leads to testicular shrinkage in men and clitoral enlargement in women), and your secondary sex characteristics (voice depth, body hair growth, man boobs, acne, etc.).

All these bad symptoms fall under the androgenic effects of steroids. The issue with steroids is that they have an anabolic-to-androgenic ratio of 1:1. This means they are just as likely to shrink your balls or enlarge your clitoris as they are to build muscle. What if you could turn on muscle building without the other androgen problems?

This is where SARMs innovate. They're far more selective than steroids, with anabolic-to-androgenic ratios starting at 3:1 and going as high as 90:1. You can still get muscle growth and fat loss, but SARMs won't give you man boobs or turn you into the bearded lady. SARMs are also legal as long as you buy them "for research purposes only." You'll notice SARMs retailers include disclaimers like "for lab research purposes only" and "not for human consumption." They do this because the substances aren't approved for human use, and they don't want to get sued. There are about a dozen SARMs in either clinical (human) trials or preclinical (animal) trials. Be careful. It's hard to find reliable suppliers, and there are many people online selling poor quality imitation SARMs. It can be difficult to source exactly what you want.

All of that said, the results I got from trying a short course of SARMs as research for this book were nothing short of unbelievable. Within six weeks I gained twenty-nine pounds of muscle without changing my workout or my diet. It happened so fast that when I was in my hotel room getting ready to go onstage at a Tony Robbins event I couldn't button a shirt that had fit just a few weeks earlier. Lots of people would probably love these results, and though I did enjoy the look for a little while, I also know that the best way to age quickly is to be either too muscular or not muscular enough. When *The New York Times* referred to me a few years prior as "almost muscular," I high-fived myself. That's actually exactly what I want to be.

Your goals might be different. If you want to be a solid wall of chiseled muscle, my hat's off to you. And if you want to join the Calorie Restriction Society (that's a real thing—it's now called the CR Society International), that's also your right. I won't judge you. Biohacking is about gaining full control of your own biology. But my goals are different, so my decisions will be different. My goal when taking the SARMs was not to build muscle, but rather to promote systemic healing and mitochondrial biogenesis. I want younger, more plentiful power plants in my cells so I can have more energy to fuel my brain and Super Human level regeneration. The SARMs I took gave me those results, and then some.

These substances can help you put on muscle very quickly, which can be lifesaving if you're seventy and suffering from muscle wasting. But if you're younger, it's possible to put on muscle faster than your body can strengthen your ligaments to support the new muscle. If you push your new muscles to the max, you have a higher chance of damaging a ligament. It's important to limit your max until your ligaments catch up to your muscles! The good news is that if you do injure yourself, some of the compounds here can give you Super Human levels of tissue regeneration.

Here is a rundown of the SARMs I've tried with mixed but powerful results:

MK-2866

With multiple published human trials under its belt, MK-2866, also known as the drug Ostarine, is one of the best-studied SARMs. Though it is weaker than many others on this list, it still has been shown to offer powerful results. In studies, Ostarine has few meaningful side effects and is very effective at building muscle. Healthy elderly men and women who took Ostarine for twelve weeks saw significant increases in lean body mass and a decrease in fat mass, and were better able to climb stairs.[8] Interestingly, these men and women also had an average decline of 11 percent in fasting blood glucose, a 17 percent reduction in insulin levels, and a 27 percent reduction in insulin resistance. This suggests that SARMs might be able to impact type 2 diabetes.

The study noted no side effects, but some people have reported short-term testosterone suppression when they take high doses of Ostarine. In these cases, testosterone rebounded to normal levels within a couple of weeks after stopping the drug. The dosage I took is far below the level that would affect testosterone levels. But there's still a risk of short-term testosterone suppression, and of course there may be other long-term side effects that we don't know about yet.

Because they're so new, dose recommendations for SARMs vary. Online communities report results when taking 15 to 20 mg of Ostarine daily for four weeks. The time of day doesn't matter. To be cautious, experienced users advise taking at least four weeks off after completing a four-week dose so your system balances out before you start another cycle. Some do a mild "post-cycle therapy" of testosterone-boosting herbs such as ashwagandha or tribulus terrestris.

LGD-4033—MUSCLE UP

Also known as Ligandrol or Anabolicum, this is another one of the better-studied SARMs. It's been through multiple human trials with interesting results. In one study, healthy men between the ages of twenty-one and fifty were broken into two groups. One group took LGD-4033 for twenty-one days, and the other took a placebo. The men who received LGD-4033 did experience a dose-dependent suppression of total testosterone, sex hormone binding globulin, high-density lipoprotein cholesterol, and triglyceride levels. These decreases were all slight—none of the men's testosterone levels fell below the normal range. And they did see a significant increase in lean body mass without a decrease in fat mass. Their hormone levels and lipids returned to baseline after the muscles grew and the treatment was discontinued.

Whether you're a man or woman, if your testosterone is already on the low side, you don't want to suppress it further. It's not worth it to potentially push your levels below the normal range. However, if you're already on bioidentical testosterone replacement, testosterone levels won't drop. Users report success taking 2 to 5 mg of LGD-4033 in a single daily oral dose for four weeks to build muscle. The higher the dose, the more muscle can develop, but the more your testosterone will dip. Many users compensate by using Clomid, a prescription drug that is normally marketed as a fertility drug for women, which helps your body recover testosterone levels faster. Once the four weeks are up, users usually wait at least a month before starting another cycle.

GW501516—EXERCISE IN A BOTTLE

GW501516 (Cardarine) isn't actually a SARM because it doesn't impact your hormone receptors, but it's often mistakenly classified as one. There have been no published human studies on this drug, but in rodents it has shown great promise as an exercise mimetic, meaning it lights up many of the same longevity-promoting genes you'd activate by exercising.[9] That alone doesn't seem to be enough to get great results, but when researchers gave mice GW501516 *and* had them exercise consistently, the results went through the roof. This combination led the mice to increase their running times by 68 percent and their distance by 70 percent while doubling their overall muscular endurance—in just five weeks. That's a Super Human (or Super Mouse?) level of performance . . . for about $50.

Another study on rodents showed that GW501516 plus exercise led to a roughly 50 percent increase in mitochondrial growth.[10] It was this study that made me want to use this compound at low doses for anti-aging purposes. A 50 percent increase in mitochondria would obviously enhance every part of my body, including the brain. The idea of having a bigger power supply for everything I do is exciting.

Of course there's a caveat. Shortly after it was classified as a performance-enhancing drug, a report came out saying that GW501516 caused cancer in lab rats. As with most substances, the devil is in the dosing. One study showed that GW501516 promoted cancer when rats took the human equivalent of 2,400 mg a day for two years straight.[11] That's about 240 times a normal dose taken every day for 104 weeks. No studies have found evidence that GW501516 causes cancer at a dose you would actually use or even at doses considerably higher than that. Plus, having better-functioning mitochondria is known to reduce cancer risk. Other rat studies report no side effects, and people in the online SARMs community report few, including no testosterone suppression.

Of course this doesn't mean there aren't side effects. We may just not know about them yet. Proceed with caution. Users report that GW501516 works best if you split it into two daily doses: 5 mg in the morning and another 5 mg in the afternoon, for a total of 10 mg a day.

SR9009

Like GW501516, SR9009 (Stenabolic) has been praised as "exercise in a pill," and in many ways it seems like the perfect supplement. In mice, it increases endurance and fat burning, decreases inflammation, and stimulates the growth of new mitochondria in muscle cells.[12] When researchers injected obese rats with SR9009, the rats lost 60 percent more weight than rats injected with a placebo, without changes to diet or exercise.

Assuming SR9009 works in humans, too, that sounds great. But the fact that the rats were injected turns out to be key. Taking SR9009 orally is pretty much useless. It has about 2 percent oral bioavailability, and your system clears that 2 percent almost immediately. This is too bad, especially because most SARMs manufacturers sell SR9009 as an oral supplement that's not suitable for injection. Unless you find injection-grade SR9009 and are willing to stick yourself a couple of times a day, you're better off spending your money on another SARM on this list.

ADDITIONAL SUPER HUMAN COMPOUNDS

Peptides are far from the only type of controversial anti-aging substance I've experimented with. Here are a few more of the most promising yet unknown anti-aging treatments out there.

HEROIN (OKAY, LOW-DOSE NALTREXONE)

Naltrexone is an opioid receptor antagonist, meaning it fits into an opiate receptor and blocks the effects of opioids. At full strength, it is used as a medication to treat alcohol and opioid drug addiction, but it has a slew of anti-aging benefits at low doses.

The first human study on low-dose Naltrexone (LDN) took place in 2007 with patients suffering from Crohn's disease. After twelve weeks of treatment, 89 percent of patients experienced a significant reduction in symptoms and 67 percent achieved full remission![13] The researchers concluded that LDN was a "novel anti-inflammatory agent in the central nervous system."[14]

Since then, low-dose Naltrexone has been studied as a treatment option for many autoimmune diseases, particularly fibromyalgia. In two separate studies, low-dose Naltrexone significantly reduced fibromyalgia pain in approximately 60 percent of participants.[15] It also seems that low-dose Naltrexone can help you avoid the Four Killers. In one study, it helped suppress tumor growth in patients with ovarian cancer.[16] And there are numerous anecdotal reports that low-dose Naltrexone suppresses tumor cell growth in B-cell lymphoma, pancreatic cancer, squamous cell carcinoma of the head and neck, and colon cancer.[17]

This information may seem new, but the idea of messing with your opioid receptors for anti-aging effects has been around since at least the Victorian era, when it was well known that heroin users lived longer and looked younger than nonusers. When Dr. Lana worked as a drug and alcohol addiction emergency room doctor in Stockholm, Sweden, the CEO of a pharmaceutical company there manufactured medical-grade heroin and sold it to a few members of the city's elite.

These people were not getting high or addicted. They used low doses once or twice a week for anti-aging purposes for more than a decade. And it's pretty telling that over twenty years, none of them tried to up their dose. All of them had visible anti-aging benefits, possibly because heroin (and many other opiates) raises levels of human growth hormone. When people found out, it led to a huge scandal with legal implications. But this was a "crime" in which there were no victims—perhaps other than the grim reaper, who was temporarily deprived because some forward-thinking people stayed young and avoided the Four Killers.

To be clear, I am not advocating for heroin use or the use of pharmaceutical opiates. Abusing heroin (and likely other opiates) actually shortens telomeres, especially in the brain,[18] and opiates are as addictive as all hell. Given that they are illegal, we have little control over what contaminants are in street drugs. I have tremendous empathy for the people whose lives have been ruined by all forms of addiction, especially the synthetic opiates on the market today that are a thousand times more active than natural opiates. We've created this situation by systematically preventing people from treating real physical pain with the most effective painkillers while simultaneously failing to treat the trauma and emotional pain behind all addiction. Chronic pain will make you age more quickly and destroy your quality of life. So will addiction.

Most of us won't be using low-dose heroin for anti-aging any time soon, but it looks like you can get an affordable anti-aging benefit from a prescription for a microdose of Naltrexone. Physicians normally prescribe 4.5 mg LDN capsules for inflammation or aging, and it has zero addiction or abuse potential. The case for using it as you age is compelling.

CARBON 60 FOR A 90 PERCENT LONGER LIFE

Another interesting compound is carbon 60, which was discovered in the 1980s when scientists realized they could form strange structures of sixty carbon atoms. These structures were incredibly stable and resembled the geodesic dome shape containing linked pentagons and hexagons originally designed by architect Buckminster Fuller. The three lead scientists later won the Nobel Prize in Chemistry for this discovery. They named the shape buckminsterfullerene in honor of Fuller, but it is more commonly known as carbon 60.

Carbon 60 is a superconductor, which may be why in studies it helps your mitochondria efficiently complete the chemical process it uses to produce energy. It also has a powerful antioxidant effect on the fats in your body.[19] It even inactivates some viruses.[20] Because it can cross the lipid bilayer membranes of cells, carbon 60 causes antioxidants to scavenge for and destroy free radicals inside of your cells, leading to powerful anti-aging effects.[21]

How powerful? In a 2012 study on rats, carbon 60 led to a 90 *percent* life-span increase. With the average human life-span at seventy-nine, carbon 60 could theoretically help the average person get to a hundred and fifty. Even if the life-span increase in humans isn't exactly the same as in rats, that's a pretty amazing statistic. The researchers conducting the study concluded that this dramatic effect on life-span was mainly due to the attenuation of age-associated increases in oxidative stress.[22]

Because of my anti-aging nonprofit work, I heard about carbon 60 in the early 2000s, before many of these studies were conducted. I ordered some from the only supplier at the time and received an unlabeled bottle in a white box. Carbon 60 always comes dissolved in oil. This one tasted like slightly rancid olive oil. But every time I took it I felt more inflamed, not less, so I threw the rest of the bottle away. I figured that this was simply one anti-aging hack that wasn't for me.

A few years later I met a biochemist and pharmaceutical designer named Ian Mitchell. He explained that I was getting my carbon 60 in oxidized olive oil, which was causing the inflammation. Carbon 60 is not patentable, so pharmaceutical companies can't make a huge profit on it and let it languish for decades. The result is that it's difficult to find a reliable source for this powerful compound. Ian's company, C360 Health, manufactures a carbon 60 product for pets, so I tried it on my thirteen-year-old dachshund, Merlin. His energy definitely perked up to the extent that I started taking the pet version myself until Ian came out with Carbon60 Plus, which is designed for humans. I noticed great improvements in energy.

I continue to take the recommended dose of about 2 teaspoons of Carbon60 Plus, which at the time of this writing costs about $25 for a six-week supply. I'll be quite happy if it provides me with even a fraction of what carbon 60 did for those rats! (Full disclosure: After interviewing Ian and extensive investigation, I became an advisor and investor in his company.)

Bottom Line

Want to heal like a deity? Do these things right now:

• Try a bioregulator peptide that will help you reduce your risk of one of the Four Killers.

• If you are suffering from an autoimmune disease or cancer, talk to your doctor about low-dose Naltrexone. This drug is available only with a prescription and it is currently prescribed to treat alcohol and opiate drug abuse. Talk to your doctor about off-label use.

• Try Carbon60 Plus, a novel and noticeable anti-aging compound.

AFTERWORD

Did you think you were going to be able to finish this book without another mythology lesson? Well, think again. It's amazing how much the ancient Greeks thought and wrote about their own quest for immortality and how similar their desires were to our own.

With that in mind, perhaps you recall the story of Tithonus, a human who was beloved by Eos, the goddess of dawn. Eos loved Tithonus so much that she begged Zeus to grant him immortality, but she was so caught up that she forgot to ask for eternal youthfulness. Zeus indeed made Tithonus immortal, but he degenerated as he aged, his hair turned white, and he lost the use of his limbs. At that point, Eos shut Tithonus up in his bedchamber, where he withered and "babbled endlessly" for eternity.

This is depressingly similar to the way humans now age, even all these centuries later. Those who manage to live for a long time usually wither and lose their faculties. And if you don't do something now to stop it, the chances are pretty high that this is exactly what's going to happen to you. By now it is clear to you that this doesn't have to be your fate. It is possible to gain energy as you age, instead of losing it. And even if you think you're too young to be concerned about aging right now, the interventions you start today will benefit your performance immediately while preventing you from aging like Tithonus in the future.

So decide right now which ones you're going to try first, and know that the rest of the technologies in this book are available to you and improving every day if and when you need them. If they're too expensive, help ramp up demand so the technologies that cost thousands of dollars today are available for pennies when you're old. If anti-aging is still a rich person's game by the time I'm a hundred and eighty, then we have failed as a species. Join me in preventing this from happening.

As I was writing this book, I celebrated my forty-sixth birthday, a number that would strike many people as depressing because in their dark vision of the future, it marks the beginning of decline. At forty-six, my best years are behind me, right?

Screw that. I know it is possible that I can make it to a hundred and eighty with my biology and my faculties still intact. And with that in mind, I blew out the candles on my (Bulletproof, of course) birthday cake thinking about the fact that I was celebrating my 25 percent birthday. I'm not middle-aged at all, and I'm so excited to accumulate and share more wisdom over the next 75 percent of my life.

Thanks to all of the information in this book, I'm not planning on plateauing or heading downhill any time soon. The dawn is not setting on my ability to make an impact. (Hear that, Eos?) In fact, I'm just getting started. And so are you.

The anti-aging technologies in this book are evolving at a very fast rate. I would be more than happy to share occasional short updates with you about how my regimen changes and how you can stay Super Human. Sign up at daveasprey.com/superhuman.

ACKNOWLEDGMENTS

There is a selfish motive for writing a book like this. There are only two ways I know to deeply understand something—you either teach it or you write a book about it, because writing is a forcing function to make you structure your knowledge. However, writing a book comes with sacrifice. It means countless late nights at the keyboard. It means less time with my wife. It means less time with my kids. For that reason, I'm acknowledging my family first—not only for giving up that time, but for supporting me so fully during the writing of this book. They are the reason that I have done everything in my power to make this book worth your time to read. I wouldn't sacrifice my own time to write it otherwise! (Read *Perennial Seller* by Ryan Holiday if you want to know more about what makes authors like me tick.)

Profound thanks to my writing partner Jodi Lipper, editor Julie Will, and agent Celeste Fine. Words do not express how impressed I am with each of your literary superpowers. Super thanks also to Anie Tazian, Beverly Hampson, and Nikki de Goey, my assistants at Bulletproof, who manage my insane calendar to make sure I hit my deadlines (mostly) and be a father, CEO, author, podcaster, and still have time to recover and practice my own anti-aging self-upgrades.

As you read in the first chapter, a book like this can exist only when it is based on the shoulders of millennia of research, including the researchers behind every single paper I referenced. While each has my gratitude, I could not name them all, and if I did, you'd stop reading. Always remember how many people are working on solving the aging problem! Special thanks to Aubrey de Grey for his pioneering work in aging and friendship for years. Thanks to Satchin Panda from the Salk Institute for major breakthroughs in mitochondrial biology and meal timing. Thanks to Mary Enig for explaining fats so well. Much appreciation for Dale Bredesen for so clearly identifying the three big causes of Alzheimer's. Thanks to Steve Fowkes for his tireless work in anti-aging and biochemistry and long-term friendship. So much appreciation for Dr. Shallenberger and Dr. Rowen for their work in ozone therapy and mitochondrial respiration. Much gratitude to T. S. Wiley and Dr. Paul Zak for their perspectives on hormones, and Dr. Klinghardt for his nearly forty years of work on toxins and biology. And a special stem cell thanks for the treatments, and the research and learning, from Dr. Harry Adelson and Dr. Amy Killen and Dr. Marcella Madera of Docere Clinics, and from Dr. Matt Cook at BioReset. Thanks to Dr. Daniel Amen, Dr. Mark Hyman, and Dr. David Perlmutter, for their groundbreaking leadership and friendship. Much appreciation for Dr. Barry Morguelan for his Chinese medicine teachings and his energy meditations, which I used while writing this book. Thanks to Jim Plante for his work on Klotho, and Ian Mitchell for the pioneering work with carbon 60 and aging. Much appreciation to Dr. Oz Garcia, Dr. Lionel Bissoon, and Dr. Philip Lee Miller, who ran my first hormone panels, and so many more fantastic human beings who have gifted me—and millions of *Bulletproof Radio* listeners—with their time.

Special thanks to a few friends who share extra business support and wisdom: Dan Scholnick, Mike Koenigs, Naveen Jain, Joe Polish, JJ Virgin, Michael Fishman, and Dan Sullivan.

NOTES

CHAPTER 1: THE FOUR KILLERS

1. Edward Giovannucci et al., "Diabetes and Cancer: A Consensus Report." *Diabetes Care* 33, no. 7 (2010): 1674–85, https://doi.org/10.2337/dc10–0666.

2. Christian Hölscher, "Diabetes as a Risk Factor for Alzheimer's Disease: Insulin Signalling Impairment in the Brain as an Alternative Model of Alzheimer's Disease," *Biochemical Society Transactions* 39, no. 4 (August 2011): 891–97, https://doi.org/10.1042/BST0390891.

3. Krishnan Bhaskaran et al., "Body-Mass Index and Risk of 22 Specific Cancers: A Population-Based Cohort Study of 5·24 Million UK Adults," *The Lancet* 384, no. 9945 (August 30, 2014): 755–65; Katrina F. Brown et al., "The Fraction of Cancer Attributable to Modifiable Risk Factors in England, Wales, Scotland, Northern Ireland, and the United Kingdom in 2015," *British Journal of Cancer* 118, no. 8 (April 2018): 1130–41.

4. Christopher J. L. Murray, Marie Ng, and Ali Mokdad, "The Vast Majority of American Adults Are Overweight or Obese, and Weight Is a Growing Problem Among US Children," Institute for Health Metrics and Evaluation (IHME), May 28, 2014, http://www.healthdata.org/news-release/vast-majority-american-adults-are-overweight-or-obese-and-weight-growing-problem-among.

5. "Inflammatory Hypothesis Confirmed: Reducing Inflammation Without Lowering Cholesterol Cuts Risk of Cardiovascular Events," Health Canal, August 27, 2017, https://www.healthcanal.com/blood-heart-circulation/heart-disease/240113-inflammatory-hypothesis-confirmed-reducing-inflammation-without-lowering-cholesterol-cuts-risk-cardiovascular-events.html.

6. University of Colorado at Boulder, "Fountain of Youth for Heart Health May Lie in the Gut: Age-Related Changes to Microbiome Fuel Vascular Decline, New Study Shows," ScienceDaily, March 19, 2019, www.sciencedaily.com/releases/2019/03/190319163527.htm.

7. Reza Nemati et al., "Deposition and Hydrolysis of Serine Dipeptide Lipids of Bacteroidetes Bacteria in Human Arteries: Relationship to Atheroscle-

rosis," *Journal of Lipid Research* 58 (October 2017): 1999–2007, https://doi.org/10.1194/jlr.M077792.

8. Thomas Meyer et al., "Attention Deficit-Hyperactivity Disorder Is Associated with Reduced Blood Pressure and Serum Vitamin D Levels: Results from the Nationwide German Health Interview and Examination Study for Children and Adolescents," *European Child & Adolescent Psychiatry* 26, no. 2 (February 2017): 165–75, https://doi.org/10.1007/s00787-016-0852-3.

9. Kevin McKeever, "Asperger Syndrome Tied to Low Cortisol Levels," HealthDay, April 2, 2009, https://consumer.healthday.com/cognitive-health-information-26/autism-news-51/asperger-syndrome-tied-to-low-cortisol-levels-625706.html.

10. Marc Yves Donath and Steven E. Shoelson, "Type 2 Diabetes as an Inflammatory Disease," *Nature Reviews Immunology* 11, no. 2 (February 2011): 98–107, https://doi.org/10.1038/nri2925.

11. University of California–San Diego, "Type 2 Diabetes: Inflammation, Not Obesity, Cause of Insulin Resistance," ScienceDaily, November 7, 2007, https://www.sciencedaily.com/releases/2007/11/071106133106.htm.

12. Yuehan Wang et al., "Association of Muscular Strength and Incidence of Type 2 Diabetes," *Mayo Clinic Proceedings* 94, no. 4 (April 2019): 643–51, https://doi.org/10.1016/j.mayocp.2018.08.037.

13. Sandra Weimer et al., "D-Glucosamine Supplementation Extends Life Span of Nematodes and of Ageing Mice," *Nature Communications* 5 (April 8, 2014): 3563, https://doi.org/10.1038/ncomms4563.

14. Richard Weindruch and Rajindar S. Sohal, "Seminars in Medicine of the Beth Israel Deaconess Medical Center. Caloric Intake and Aging," *New England Journal of Medicine* 337, no. 14 (October 2, 1997): 986–94, https://doi.org/10.1056/NEJM199710023371407.

15. "D-Glucosamine as an Example of Calorie Restriction Mimetic Research," Fight Aging!, April 8, 2014, https://www.fightaging.org/archives/2014/04/d-glucosamine-as-an-example-of-calorie-restriction-mimetic-research/.

16. Karen W. Della Corte et al., "Effect of Dietary Sugar Intake on Biomarkers of Subclinical Inflammation: A Systematic Review and Meta-Analysis of Intervention Studies," *Nutrients* 10, no. 5 (2018): 606, https://doi.org/10.3390/nu10050606.

17. Santosh Kumar Singh, "Post-Prandial Hyperglycemia," *Indian Journal of Endocrinology and Metabolism* 16, no. 8 (December 2012): 245–47, https://doi.org/10.4103/2230-8210.104051.

18. Federation of American Societies for Experimental Biology, "Scientists Remove Amyloid Plaques from Brains of Live Animals with Alzheimer's Disease," *ScienceDaily*, www.sciencedaily.com/releases/2009/10/091015091602.htm (accessed July 16, 2019).

19. "41 Percent of Americans Will Get Cancer," UPI Health News, May 6, 2010, https://www.upi.com/41-percent-of-Americans-will-get-cancer/75711273192042/.

20. Lisa M. Coussens and Zena Werb, "Inflammation and Cancer," *Nature* 420, no. 6917 (2002): 860–67, https://doi.org/10.1038/nature01322.

CHAPTER 2: THE SEVEN PILLARS OF AGING

1. Helen Karakelides and K. Sreekumaran Nair, "Sarcopenia of Aging and Its Metabolic Impact," *Current Topics in Developmental Biology* 68 (2005): 123–48, https://doi.org/10.1016/S0070-2153(05)68005-2.

2. Elena Volpi, Reza Nazemi, and Satoshi Fujita, "Muscle Tissue Changes with Aging," *Current Opinion in Clinical Nutrition and Metabolic Care* 7, no. 4 (2004): 405–10, https://doi.org/10.1097/01.mco.0000134362.76653.b2.

3. James Golomb et al., "Hippocampal Atrophy in Normal Aging. An Association with Recent Memory Impairment," *Archives of Neurology* 50, no. 9 (September 1993): 967–73, https://doi.org/10.1001/archneur.1993.00540090066012.

4. Martin Stimpfel, Nina Jancar, and Irma Virant-Klun, "New Challenge: Mitochondrial Epigenetics?," *Stem Cell Reviews and Reports* 14, no. 1 (February 2018): 13–26, https://doi.org/10.1007/s12015-017-9771-z.

5. James L. Kirkland and Tamara Tchkonia, "Cellular Senescence: A Translational Perspective," *EBioMedicine* 21 (July 2017): 21–28, https://doi.org/10.1016/j.ebiom.2017.04.013.

6. Viktor I. Korolchuk et al., "Mitochondria in Cell Senescence: Is Mitophagy the Weakest Link?," *EBioMedicine* 21 (July 2017): 7–13, https://doi.org/10.1016/j.ebiom.2017.03.020.

7. Okhee Jeon et al., "Senescent Cells and Osteoarthritis: A Painful Connection," *Journal of Clinical Investigation* 128, no. 4 (April 2, 2018): 1229–37, https://doi.org/10.1172/JCI95147.

8. Derek M. Huffman, Marissa J. Schafer, and Nathan K. LeBrasseur, "Energetic Interventions for Healthspan and Resiliency with Aging," *Experimental Gerontology* 86 (December 15, 2016): 73–83, https://doi.org/10.1016/j.exger.2016.05.012.

9. Christian A. Bannister et al., "Can People with Type 2 Diabetes Live Longer Than Those Without? A Comparison of Mortality in People Initiated with Metformin or Sulphonylurea Monotherapy and Matched, Non-Diabetic Controls," *Diabetes, Obesity and Metabolism* 16, no. 11 (November 2014): 1165–73, https://doi.org/10.1111/dom.12354.

10. Agnieszka Śmieszek et al., "Antioxidant and Anti-Senescence Effect of Metformin on Mouse Olfactory Ensheathing Cells (mOECs) May Be Associated with Increased Brain-Derived Neurotrophic Factor Levels—An Ex Vivo Study," *International Journal of Molecular Sciences* 18, no. 4 (2017): 872, https://doi.org/10.3390/ijms18040872.

11. Rong Wang et al., "Rapamycin Inhibits the Secretory Phenotype of Senescent Cells by a Nrf2-Independent Mechanism," *Aging Cell* 16, no. 3 (June 2017): 564–74, https://doi.org/10.1111/acel.12587.

12. "Animal Data Shows Fisetin to Be a Surprisingly Effective Senolytic," Fight Aging!, October 3, 2018, https://www.fightaging.org/archives/2018/10/animal-data-shows-fisetin-to-be-a-surprisingly-effective-senolytic/.

13. Pamela Maher, "How Fisetin Reduces the Impact of Age and Disease on CNS Function," *Frontiers in Bioscience (Scholar Edition)* 7 (June 1, 2015): 58–82, https://www.ncbi.nlm.nih.gov/pubmed/25961687.

14. Kashmira Gander, "Secret of Longevity Could Be Found in Traditional Japanese Plant that Appears to Slow Aging," *Newsweek*, February 20, 2019, https://www.newsweek.com/anti-aging-longevity-japanese-plant-1336734.

15. "Uncovering the Senolytic Mechanism of Piperlongumine," Fight Aging!, May 21, 2018, https://www.fightaging.org/archives/2018/05/uncovering-the-senolytic-mechanism-of-piperlongumine/.

16. Yin-Ju Chen et al., "Piperlongumine Inhibits Cancer Stem Cell Properties and Regulates Multiple Malignant Phenotypes in Oral Cancer," *Oncology Letters* 15, no. 2 (February 2018): 1789–98, https://doi.org/10.3892/ol.2017.7486.

17. Fernanda de Lima Moreira et al., "Metabolic Profile and Safety of Piperlongumine," *Nature Scientific Reports* 6 (September 29, 2016): article no. 33646, https://www.nature.com/articles/srep33646.

18. Alan R. Gaby, "Adverse Effects of Dietary Fructose," *Alternative Medicine Review* 10, no. 4 (December 2005): 294–306, http://www.ncbi.nlm.nih.gov/pubmed/16366738.

19. Matthew Streeter et al., "Identification of Glucosepane Cross-Link Breaking Enzymes," *Diabetes* 67, no. S1 (July 2018): 1229-P, https://doi.org/10.2337/db18-1229-P.

20. Xu Wang et al., "Insulin Deficiency Exacerbates Cerebral Amyloidosis and Behavioral Deficits in an Alzheimer Transgenic Mouse Model," *Molecular Neurodegeneration* 5 (2010): 46, https://doi.org/10.1186/1750-1326-5-46.

21. Jordan Lite, "Vitamin D Deficiency Soars in the U.S., Study Says," *Scientific American*, March 23, 2009, https://www.scientificamerican.com/article/vitamin-d-deficiency-united-states/.

22. Society for Neuroscience, "Staving Off Alzheimer's Disease with the Right Diet, Prescriptions," ScienceDaily, November 13, 2007, https://www.sciencedaily.com/releases/2007/11/071107211036.htm.

23. Gabriella Notarachille et al., "Heavy Metals Toxicity: Effect of Cadmium Ions on Amyloid Beta Protein 1-42. Possible Implications for Alzheimer's Disease," *Biometals* 27, no. 2 (April 2014): 371–88, https://doi.org/10.1007/s10534-014-9719-6.

24. Paul B. Tchounwou et al., "Heavy Metal Toxicity and the Environment," in *Molecular, Clinical and Environmental Toxicology*, Experientia Sup-

plementum, vol. 101, ed. Andrea Luch (Basel, CH: Springer, 2012): 133–64.

25. Elena A. Belyaeva et al., "Mitochondria as an Important Target in Heavy Metal Toxicity in Rat Hepatoma AS-30D Cells," *Toxicology and Applied Pharmacology* 231, no. 1 (August 15, 2008): 34–42, https://doi.org/10.1016/j.taap.2008.03.017.

26. Varun Parkash Singh et al., "Advanced Glycation End Products and Diabetic Complications," *The Korean Journal of Physiology & Pharmacology* 18, no. 1 (2014): 1–14, https://doi.org/10.4196/kjpp.2014.18.1.1.

27. David P. Turner, "Advanced Glycation End-Products: A Biological Consequence of Lifestyle Contributing to Cancer Disparity," *Cancer Research* 75, no. 10 (May 2015): 1925–29, https://doi.org/10.1158/0008-5472.CAN-15-0169.

28. Melpomeni Peppa and Sotirios A. Raptis, "Advanced Glycation End Products and Cardiovascular Disease," *Current Diabetes Reviews* 4, no. 2 (May 2008): 92–100, https://www.ncbi.nlm.nih.gov/pubmed/18473756.

29. Nobuyuki Sasaki et al., "Advanced Glycation End Products in Alzheimer's Disease and Other Neurodegenerative Diseases," *American Journal of Pathology* 153, no. 4 (October 1998): 1149–55, https://doi.org/10.1016/S0002-9440(10)65659-3.

30. *The BMJ*, "Fried Food Linked to Heightened Risk of Early Death Among Older US Women: Fried Chicken and Fried Fish in Particular Seem to Be Associated with Higher Risk of Death," ScienceDaily, January 23, 2019, https://www.sciencedaily.com/releases/2019/01/190123191637.htm.

31. "Hayflick Limit," ScienceDirect, https://www.sciencedirect.com/topics/medicine-and-dentistry/hayflick-limit.

32. Pim van der Harst et al., "Telomere Length of Circulating Leukocytes Is Decreased in Patients with Chronic Heart Failure," *Journal of the American College of Cardiology* 49, no. 13 (April 3, 2007): 1459–64, https://doi.org/10.1016/j.jacc.2007.01.027; Annette L. Fitzpatrick et al., "Leukocyte Telomere Length and Cardiovascular Disease in the Cardiovascular Health Study," *American Journal of Epidemiology* 165, no. 1 (January 1, 2007): 14–21, https://doi.org/10.1093/aje/kwj346; Robert Y. L. Zee et al., "Association of Shorter Mean Telomere Length with Risk of Incident Myocardial Infarction: A Prospective, Nested Case-Control Approach," *Clinica Chemica Acta* 403, no. 1–2, (May 2009): 139–41, https://doi.org/10.1016/j.cca.2009.02.004.

33. Monica McGrath et al., "Telomere Length, Cigarette Smoking, and Bladder Cancer Risk in Men and Women," *Cancer Epidemiology, Biomarkers & Prevention* 16, no. 4 (April 2007): 815–19, https://doi.org/10.1158/1055-9965.EPI-06-0961.

34. Mike J. Sampson et al., "Monocyte Telomere Shortening and Oxidative DNA Damage in Type 2 Diabetes," *Diabetes Care* 29, no. 2 (February 2006): 283–89, https://doi.org/10.2337/diacare.29.02.06.dc05-1715.

35. Ana M. Valdes et al., "Telomere Length in Leukocytes Correlates with Bone Mineral Density and Is Shorter in Women with Osteoporosis," *Osteoporosis International* 18, no. 9 (September 2007): 1203–10, https://doi .org/10.1007/s00198-007-0357-5.

36. Masood A. Shammas, "Telomeres, Lifestyle, Cancer, and Aging," *Current Opinion in Clinical Nutrition and Metabolic Care* 14, no.1 (January 2011): 28–34, https://doi.org/10.1097/MCO.0b013e32834121b1.

37. Richard M. Cawthon et al., "Association Between Telomere Length in Blood and Mortality in People Aged 60 Years or Older," *The Lancet* 361, no. 9355 (February 1, 2003): 393–95, https://doi.org/10.1016/S0140 -6736(03)12384-7.

38. Elissa S. Epel, "Accelerated Telomere Shortening in Response to Life Stress," *Proceedings of the National Academy of Science of the USA* 101, no. 49 (December 7, 2004): 17312–15, https://doi.org/10.1073/pnas.040716210.

39. Gretchen Reynolds, "Phys Ed: How Exercising Keeps Your Cells Young," *New York Times* Well, January 27, 2010, https://well.blogs.nytimes.com/2010/01/27 /phys-ed-how-exercising-keeps-your-cells-young/?scp=1&sq=how%20 exercising%20keeps%20your%20cells%20young&st=cse.

40. Angela R. Starkweather, "The Effects of Exercise on Perceived Stress and IL-6 Levels Among Older Adults," *Biological Research for Nursing* 8, no. 3 (January 2007): 186–94, https://www.ncbi.nlm.nih.gov/pubmed/17172317.

41. Vladimir N. Anisimov et al., "Effect of Epitalon on Biomarkers of Aging, Life Span and Spontaneous Tumor Incidence in Female Swiss-derived SHR Mice," *Biogerontology* 4, no. 4 (2003): 193–202, https://doi .org/10.1023/A:1025114230714.

42. George Kossoy et al., "Epitalon and Colon Carcinogenesis in Rats: Proliferative Activity and Apoptosis in Colon Tumors," *International Journal of Molecular Medicine* 12, no. 4 (October 2003): 473–75, https://doi.org /10.3892/ijmm.12.4.473.

43. Brenda Molgora et al., "Functional Assessment of Pharmacological Telomerase Activators in Human T Cells," *Cells* 2, no. 1 (March 2013): 57–66, https://doi.org/10.3390/cells2010057.

CHAPTER 3: FOOD IS AN ANTI-AGING DRUG

1. Kyung-Ah Kim et al., "Gut Microbiota Lipopolysaccharide Accelerates Inflamm-Aging in Mice," *BMC Microbiology* 16, no. 1 (2016): 9, https:// doi.org/10.1186/s12866-016-0625-7; Yong-Fei Zhao et al., "The Synergy of Aging and LPS Exposure in a Mouse Model of Parkinson's Disease," *Aging and Disease* 9, no. 5 (2018): 785–97, https://doi.org/10.14336/AD .2017.1028.

2. Ki Wung Chung et al., "Age-Related Sensitivity to Endotoxin-Induced Liver Inflammation: Implication of Inflammasome/IL-1β for Steatohepatitis," *Aging Cell* 14, no. 4 (April 2015): 526, fig. 1, https://doi.org/10.1111 /acel.12305.

3. Caria Sategna-Guidetti et al., "Autoimmune Thyroid Disease and Co-eliac Disease," *European Journal of Gastroenterology & Hepatology* 10, no. 11 (November 1998): 927–31, http://www.ncbi.nlm.nih.gov/pubmed /9872614.

4. A. J. Batchelor and Juliet E. Compston, "Reduced Plasma Half-Life of Radio-Labelled 25-Hydroxyvitamin D_3 in Subjects Receiving a High-Fibre Diet," *British Journal of Nutrition* 49, no. 2 (March 1983): 213–16, https ://doi.org/10.1079/BJN19830027.

5. Siriporn Thongprakaisang et al., "Glyphosate Induces Human Breast Cancer Cells Growth via Estrogen Receptors," *Food and Chemical Toxicology* 59 (September 2013): 129–36, https://doi.org/10.1016/j.fct.2013.05.057.

6. Francisco Peixoto, "Comparative Effects of the Roundup and Glyphosate on Mitochondrial Oxidative Phosphorylation," *Chemosphere* 61, no. 8 (December 2005): 1115–22, https://doi.org/10.1016/j.chemosphere.2005 .03.044.

7. Anthony Samsel and Stephanie Seneff, "Glyphosate, Pathways to Modern Diseases IV: Cancer and Related Pathologies," *Journal of Biological Physics and Chemistry* 15 (2015): 121–59, https://doi.org/10.4024/11SA15R .jbpc.15.03.

8. Stephanie Seneff and Laura F. Orlando, "Glyphosate Substitution for Glycine During Protein Synthesis as a Causal Factor in Mesoamerican Nephropathy," *Journal of Environmental & Analytical Toxicology* 8, no. 1 (2018): 541, https://doi.org/10.4172/2161-0525.1000541.

9. James H. O'Keefe, Neil M. Gheewala, and Joan O. O'Keefe, "Dietary Strategies for Improving Post-Prandial Glucose, Lipids, Inflammation, and Cardiovascular Health," *Journal of the American College of Cardiology* 51, no. 3 (January 22, 2008): 249–55, https://doi.org/10.1016/j.jacc.2007.10.016.

10. Başar Altınterim, "Anti-Throid Effects of PUFAs (Polyunsaturated Fats) and Herbs," *Trakya University Journal of Natural Sciences* 13, no. 2 (2012): 87–94, https://www.researchgate.net/publication/268515453_anti -throid_effects_of_pufas_polyunsaturated_fats_and_herbs.

11. Morgan E. Levine et al., "Low Protein Intake Is Associated with a Major Reduction in IGF-1, Cancer, and Overall Mortality in the 65 and Younger but Not Older Population," *Cell Metabolism* 19, no. 3 (March 4, 2014): 407–17, https://doi.org/10.1016/j.cmet.2014.02.006.

12. John F. Trepanowski et al., "Impact of Caloric and Dietary Restriction Regimens on Markers of Health and Longevity in Humans and Animals: A Summary of Available Findings," *Nutrition Journal* 10 (October 7, 2011): 107, https://doi.org/10.1186/1475-2891-10-107.

13. Okinawa Institute of Science and Technology (OIST) Graduate University, "Fasting Ramps Up Human Metabolism, Study Shows," ScienceDaily, January 31, 2019, https://www.sciencedaily.com/releases/2019/01/190131113934 .htm.

14. Mehrdad Alirezaei et al., "Short-Term Fasting Induces Profound Neuronal

Autophagy," *Autophagy* 6, no. 6 (August 2010): 702–10, https://doi.org/10.4161/auto.6.6.12376.

15. Behnam Sadeghirad et al., "Islamic Fasting and Weight Loss: A Systematic Review and Meta-Analysis," *Public Health Nutrition* 17, no. 2 (February 1, 2014): 396–406, https://doi.org/10.1017/S1368980012005046.

16. Mark P. Mattson, Wenzhen Duan, and Zhihong Guo, "Meal Size and Frequency Affect Neuronal Plasticity and Vulnerability to Disease: Cellular and Molecular Mechanisms," *Journal of Neurochemistry* 84, no. 3 (February 2003): 417–31, https://doi.org/10.1046/j.1471-4159.2003.01586.x.

17. Gerrit van Meer, Dennis R. Voelker, and Gerald W. Feigenson, "Membrane Lipids: Where They Are and How They Behave," *Nature Reviews Molecular Cell Biology* 9, no. 2 (February 2008): 112–24, https://doi.org/10.1038/nrm2330.

18. Vincent Rioux, "Fatty Acid Acylation of Proteins: Specific Roles for Palmitic, Myristic and Caprylic Acids," *OCL* 23, no. 3 (May–June 2016): D304, https://doi.org/10.1051/ocl/2015070.

19. Elisa Parra-Ortiz et al., "Effects of Oxidation on the Physicochemical Properties of Polyunsaturated Lipid Membranes," *Journal of Colloid and Interface Science* 538 (March 7, 2019): 404–19, https://doi.org/10.1016/j.jcis.2018.12.007.

20. National Institutes of Health, Office of Dietary Supplements, "Omega-3 Fatty Acids: Fact Sheet for Health Professionals," U.S. Department of Health and Human Services, last modified November 21, 2018, https://ods.od.nih.gov/factsheets/Omega3FattyAcids-HealthProfessional/.

21. Neal Simonsen et al., "Adipose Tissue Omega-3 and Omega-6 Fatty Acid Content and Breast Cancer in the EURAMIC Study," *American Journal of Epidemiology* 147, no. 4 (February 15, 1998): 342–52, https://doi.org/10.1093/oxfordjournals.aje.a009456; Sanjoy Ghosh, Elizabeth M. Novak, and Sheila M. Innis, "Cardiac Proinflammatory Pathways Are Altered with Different Dietary n-6 Linoleic to n-3 Alpha-Linolenic Acid Ratios in Normal, Fat-Fed Pigs," *American Journal of Physiology: Heart and Circulatory Physiology* 293, no. 5 (November 2007): H2919–27, https://doi.org/10.1152/ajpheart.00324.2007; Urmila Nair, Helmut Bartsch, and Jagadeesan Nair, "Lipid Peroxidation-Induced DNA Damage in Cancer-Prone Inflammatory Diseases: A Review of Published Adduct Types and Levels in Humans," *Free Radical Biology & Medicine* 43, no. 8 (October 2007): 1109–20, https://doi.org/10.1016/j.freeradbiomed.2007.07.012; Véronique Chajès and Philippe Bougnoux, "Omega-6/Omega-3 Polyunsaturated Fatty Acid Ratio and Cancer," in *Omega 6/Omega 3 Fatty Acid Ratio: The Scientific Evidence*, World Review of Nutrition and Dietetics, vol. 92, ed. Artemis P. Simopoulos and Leslie G. Cleland (Basel, CH: Karger, 2003), 133–51; Emily Sonestedt et al., "Do Both Heterocyclic Amines and Omega-6 Polyunsaturated Fatty Acids Contribute to the Incidence of Breast Cancer in Postmenopausal Women of the Malmö Diet and

Cancer Cohort?," *International Journal of Cancer* 123, no. 7 (October 1, 2008): 1637–43, https://doi.org/10.1002/ijc.23394.

22. Juhee Song et al., "Analysis of *Trans* Fat in Edible Oils with Cooking Process," *Toxicological Research* 31, no. 3 (September 2015): 307–12, https://doi.org/10.5487/TR.2015.31.3.307.

23. Camille Vandenberghe et al., "Tricaprylin Alone Increases Plasma Ketone Response More Than Coconut Oil or Other Medium-Chain Triglycerides: An Acute Crossover Study in Healthy Adults," *Current Developments in Nutrition* 1, no. 4, (April 1, 2017): e000257, https://doi.org/10.3945/cdn.116.000257.

24. Arturo Solis Herrera and Paola E. Solis Arias, "Einstein Cosmological Constant, the Cell, and the Intrinsic Property of Melanin to Split and Re-Form the Water Molecule," *MOJ Cell Science & Report* 1, no. 2 (August 27, 2014): 46–51, https://doi.org/10.15406/mojcsr.2014.01.00011.

25. Ana S. P. Moreira et al., "Coffee Melanoidins: Structures, Mechanisms of Formation and Potential Health Impacts," *Food & Function* 3, no. 9 (September 2012): 903–15, https://doi.org/10.1039/c2fo30048f.

CHAPTER 4: SLEEP OR DIE

1. Matthew P. Walker et al., "Practice with Sleep Makes Perfect: Sleep-Dependent Motor Skill Learning," *Neuron* 35, no. 1 (July 2002): 205–11, https://doi.org/10.1016/S0896-6273(02)00746-8.

2. Ullrich Wagner et al., "Sleep Inspires Insight," *Nature* 247, no. 6972 (January 22, 2004): 352–55, https://doi.org/10.1038/nature02223.

3. Margaret Altemus et al., "Stress-Induced Changes in Skin Barrier Function in Healthy Women," *Journal of Investigative Dermatology* 117, no. 2 (August 2001): 309–17, https://doi.org/10.1046/j.1523-1747.2001.01373.x.

4. Philippa J. Carter et al., "Longitudinal Analysis of Sleep in Relation to BMI and Body Fat in Children: The FLAME Study," *BMJ* 342 (May 26, 2011): d2712, https://doi.org/10.1136/bmj.d2712.

5. Josephine Arendt, "Shift Work: Coping with the Biological Clock," *Occupational Medicine* 60, no. 1 (January 2010): 10–20, https://doi.org/10.1093/occmed/kqp162.

6. Guglielmo Beccuti and Silvana Pannain, "Sleep and Obesity," *Current Opinion in Clinical Nutrition & Metabolic Care* 14, no. 4 (July 2011): 402–12, https://doi.org/10.1097/MCO.0b013e3283479109.

7. Lulu Xie et al., "Sleep Drives Metabolite Clearance from the Adult Brain," *Science* 342, no. 6156 (October 18, 2013): 373–77, https://doi.org/10.1126/science.1241224.

8. National Institutes of Health, "Sleep Deprivation Increases Alzheimer's Protein," NIH Research Matters, April 24, 2018, https://www.nih.gov/news-events/nih-research-matters/sleep-deprivation-increases-alzheimers-protein.

9. Hedok Lee et al., "The Effect of Body Posture on Brain Glymphatic Transport," *The Journal of Neuroscience* 34, no. 31 (August 5, 2015): 11034–44, https://doi.org/10.1523/JNEUROSCI.1625-15.2015.

10. Masatoshi Fujita et al., "Effects of Posture on Sympathetic Nervous Modulation in Patients with Chronic Heart Failure," *The Lancet* 356, no. 9244 (November 25, 2000): 1822–23, https://doi.org/10.1016/S0140-6736(00)03240-2.

11. Ryan J. Ramezani and Peter W. Stacpoole, "Sleep Disorders Associated with Primary Mitochondrial Diseases," *Journal of Clinical Sleep Medicine: JCSM* 10, no. 11 (November 15, 2014): 1233–39, https://doi.org/10.5664/jcsm.4212.

12. Wendy M. Troxel et al., "Sleep Symptoms Predict the Development of the Metabolic Syndrome," *Sleep* 33, no. 12 (December 2010): 1633–40, https://doi.org/10.1093/sleep/33.12.1633.

13. Daniel F. Kripke et al., "Mortality Related to Actigraphic Long and Short Sleep," *Sleep Medicine* 12, no. 1 (January 2011): 28–33, https://www.ncbi.nlm.nih.gov/pubmed/11825133.

14. Joel H. Benington and H. Craig Heller, "Restoration of Brain Energy Metabolism as the Function of Sleep," *Progress in Neurobiology* 45, no. 4 (March 1995): 347–60, https://doi.org/10.1016/0301-0082(94)00057-O.

15. Scott A. Cairney et al., "Mechanisms of Memory Retrieval in Slow-Wave Sleep," *Sleep* 40, no. 9 (September 2017): zsx114, https://doi.org/10.1093/sleep/zsx114.

16. Scott A. Cairney et al., "Complementary Roles of Slow-Wave Sleep and Rapid Eye Movement Sleep in Emotional Memory Consolidation," *Cerebral Cortex* 25, no. 6 (June 2015): 1565–75, https://doi.org/10.1093/cercor/bht349.

17. Judith A. Floyd et al., "Changes in REM-Sleep Percentage over the Adult Lifespan," *Sleep* 30, no. 7 (July 1, 2007): 829–36, https://doi.org/10.1093/sleep/30.7.829.

18. "How Many Hours of Deep Sleep Does One Need?," New Health Advisor, https://www.newhealthadvisor.com/How-Much-Deep-Sleep-Do-You-Need.html.

19. "Sleep Restriction May Reduce Heart Rate Variability," Medscape, June 15, 2007, https://www.medscape.com/viewarticle/558331.

20. J. Gouin et al., "Heart Rate Variability Predicts Sleep Efficiency," *Sleep Medicine* 14, no. 1 (December 2013): e142, https://doi.org/10.1016/j.sleep.2013.11.321.

21. Marcello Massimini et al., "Triggering Sleep Slow Waves by Transcranial Magnetic Stimulation," *Proceedings of the National Academy of Sciences of the USA* 104, no. 20 (May 15, 2007): 8496–501, https://doi.org/10.1073/pnas.0702495104.

22. Giulio Tononi et al., "Enhancing Sleep Slow Waves with Natural Stimuli," *Medicamundi* 54, no. 2 (January 2010): 82–88, https://www.researchgate.net/publication/279545240_Enhancing_sleep_slow_waves_with_natural_stimuli.

23. Hong-Viet V. Ngo et al., "Auditory Closed Loop Stimulation of the Sleep Slow Oscillation Enhances Memory," *Neuron* 78, no. 3 (May 8, 2013): P545–553, https://doi.org/10.1016/j.neuron.2013.03.006; Luciana Besedovsky et al., "Auditory Closed-Loop Stimulation of EEG Slow Oscillations Strengthens Sleep and Signs of Its Immune-Supportive Function," *Nature Communications* 8, no. 1 (2017): 1984, https://doi.org/10.1038/s41467-017-02170-3.

24. Robert E. Strong et al., "Narrow-Band Blue-Light Treatment of Seasonal Affective Disorder in Adults and the Influence of Additional Nonseasonal Symptoms," *Depression and Anxiety* 26, no. 3 (2009): 273–78, https://doi.org/10.1002/da.20538.

25. Gianluca Tosini, Ian Ferguson, and Kazuo Tsubota, "Effects of Blue Light on the Circadian System and Eye Physiology," *Molecular Vision* 22 (January 24, 2016): 61–72, https://www.ncbi.nlm.nih.gov/pubmed/26900325; Anne-Marie Chang et al., "Evening Use of Light-Emitting eReaders Negatively Affects Sleep, Circadian Timing, and Next-Morning Alertness," *Proceedings of the National Academy of Sciences of the USA* 112, no. 4 (January 27, 2015): 1232–37, https://doi.org/10.1073/pnas.1418490112.

26. Tosini, Ferguson, and Tsubota, "Effects."

27. Chang et al., "Evening Use."

28. Karine Spiegel et al., "Effects of Poor and Short Sleep on Glucose Metabolism and Obesity Risk," *Nature Reviews Endocrinology* 5, no. 5 (2009): 253–61, https://doi.org/10.1038/nrendo.2009.23.

29. Ariadna Garcia-Saenz et al., "Evaluating the Association Between Artificial Light-at-Night Exposure and Breast and Prostate Cancer Risk in Spain (MCC-Spain Study)," *Environmental Health Perspectives* 126, no. 4 (April 23, 2018): 047011, https://doi.org/10.1289/EHP1837.

30. Aziz Sancar et al., "Circadian Clock Control of the Cellular Response to DNA Damage," *FEBS Letters* 584, no. 12 (June 18, 2010): 2618–25, https://doi.org/10.1016/j.febslet.2010.03.017.

31. Tosini, Ferguson, and Tsubota, "Effects."

32. Bright Focus Foundation, "Age-Related Macular Degeneration: Facts and Figures," last modified January 5, 2016, https://www.brightfocus.org/macular/article/age-related-macular-facts-figures.

33. Edward Loane et al., "Transport and Retinal Capture of Lutein and Zeaxanthin with Reference to Age-Related Macular Degeneration," *Survey of Ophthalmology* 53, no. 1 (January–February 2008): 68–81, https://doi.org/10.1016/j.survophthal.2007.10.008; Le Ma et al., "Effect of Lutein and

Zeaxanthin on Macular Pigment and Visual Function in Patients with Early Age-Related Macular Degeneration," *Ophthalmology* 119, no. 11 (November 2012): 2290–97, https://doi.org/10.1016/j.ophtha.2012.06.014.

CHAPTER 5: USING LIGHT TO GAIN SUPER POWERS

1. Ya Li et al., "Melatonin for the Prevention and Treatment of Cancer," *Oncotarget* 8, no. 24 (June 2017): 39896–921, https://doi.org/10.18632/oncotarget.16379.

2. Bhagyesh R. Sarode et al., "Light Control of Insulin Release and Blood Glucose Using an Injectable Photoactivated Depot," *Molecular Pharmacology* 13, no. 11 (November 7, 2016): 3835–41, https://doi.org/10.1021/acs.molpharmaceut.6b00633; Marla Paul, "Exposure to Bright Light May Alter Blood Sugar," Futurity, May 19, 2016, https://www.futurity.org/bright-light-metabolism-1166262–2/.

3. Nataliya A. Rybnikova, A. Haim, and Boris A. Portnov, "Does Artificial Light-at-Night Exposure Contribute to the Worldwide Obesity Pandemic?," *International Journal of Obesity* 40, no. 5 (May 2016): 815–23, https://doi.org/10.1038/ijo.2015.255.

4. Bernard F. Godley et al., "Blue Light Induces Mitochondrial DNA Damage and Free Radical Production in Epithelial Cells," *The Journal of Biological Chemistry* 280, no. 22 (June 3, 2005): 21061–66, https://doi.org/10.1074/jbc.M502194200.

5. Hajime Ishii et al., "Seasonal Variation of Glycemic Control in Type-2 Diabetic Patients," *Diabetes Care* 24, no. 8 (August 2001): 1503, https://doi.org/10.2337/diacare.24.8.1503.

6. Pelle G. Lindqvist, Håkan Olsson, and Mona Landin-Olsson, "Are Active Sun Exposure Habits Related to Lowering Risk of Type 2 Diabetes Mellitus in Women, a Prospective Cohort Study?," *Diabetes Research and Clinical Practice* 90, no. 1 (October 2010): 109–14, https://doi.org/10.1016/j.diabres.2010.06.007.

7. Sian Geldenhuys et al., "Ultraviolet Radiation Suppresses Obesity and Symptoms of Metabolic Syndrome Independently of Vitamin D in Mice Fed a High-Fat Diet," *Diabetes* 63, no. 11 (November 2011): 3759–69, https://doi.org/10.2337/db13-1675.

8. Daniel Barolet, François Christiaens, and Michael R. Hamblin, "Infrared and Skin: Friend or Foe," *Journal of Photochemistry and Photobiology B: Biology* 155 (February 2016): 78–85, https://doi.org/10.1016/j.jphotobiol.2015.12.014.

9. Pelle G. Lindqvist et al., "Avoidance of Sun Exposure as a Risk Factor for Major Causes of Death: A Competing Risk Analysis of the Melanoma in Southern Sweden Cohort," *Journal of Internal Medicine* 280, no. 4 (October 2016): 375–87, https://doi.org/10.1111/joim.12496.

10. Douglas Main, "Why Insect Populations Are Plummeting—and Why It Matters," *National Geographic*, February 14, 2019, https://www.national

geographic.com/animals/2019/02/why-insect-populations-are-plummeting
-and-why-it-matters/.

11. Cleber Ferraresi, Michael R. Hamblin, and Nivaldo A. Parizotto, "Low-Level Laser (Light) Therapy (LLLT) on Muscle Tissue: Performance, Fatigue and Repair Benefited by the Power of Light," *Photonics & Lasers in Medicine* 1, no. 4 (November 1, 2012): 267–86, https://doi.org/10.1515/plm-2012–0032.

12. Lilach Gavish et al., "Low Level Laser Irradiation Stimulates Mitochondrial Membrane Potential and Disperses Subnuclear Promyelocytic Leukemia Protein," *Lasers in Surgery and Medicine* 35, no. 5 (December 2004): 369–76, https://doi.org/10.1002/lsm.20108.

13. Pinar Avci et al., "Low-Level Laser (Light) Therapy (LLLT) in Skin: Stimulating, Healing, Restoring," *Seminars in Cutaneous Medicine and Surgery* 32, no.1 (2013): 41–52, https://www.ncbi.nlm.nih.gov/pubmed/24049929.

14. Shang-Ru Tsai et al., "Low-Level Light Therapy Potentiates NPe6-Mediated Photodynamic Therapy in a Human Osteosarcoma Cell Line via Increased ATP," *Photodiagnosis and Photodynamic Therapy* 12, no. 1 (March 2015): 123–30, https://doi.org/10.1016/j.pdpdt.2014.10.009.

15. Ulrike H. Mitchell and Gary L. Mack, "Low-Level Laser Treatment with Near-Infrared Light Increases Venous Nitric Oxide Levels Acutely: A Single-Blind, Randomized Clinical Trial of Efficacy," *American Journal of Physical Medicine & Rehabilitation* 92, no. 2 (February 2013): 151–56, https://doi.org/10.1097/PHM.0b013e318269d70a.

16. Ferraresi, Hamblin, and Parizotto, "Low-Level Laser (Light) Therapy."

17. Fernando José de Lima, Fabiano Timbó Barbosa, and Célio Fernando de Sousa-Rodrigues, "Use Alone or in Combination of Red and Infrared Laser in Skin Wounds," *Journal of Lasers in Medical Sciences* 5, no. 2 (2014): 51–57, https://www.ncbi.nlm.nih.gov/pmc/articles/PMC4291816/.

18. Ivayla I. Geneva, "Photobiomodulation for the Treatment of Retinal Diseases: A Review," *International Journal of Ophthalmology* 9, no.1 (January 2016): 145–52, https://doi.org/10.18240/ijo.2016.01.24.

19. Stephen J. Genuis et al., "Blood, Urine, and Sweat (BUS) Study: Monitoring and Elimination of Bioaccumulated Toxic Elements," *Archives of Environmental Contamination and Toxicology* 61, no. 2 (August 2011): 344–57, https://doi.org/10.1007/s00244-010-9611-5.

20. Hisashi Naito et al., "Heat Stress Attenuates Skeletal Muscle Atrophy in Hindlimb-Unweighted Rats," *Journal of Applied Physiology* 88, no. 1 (January 2000): 359–63, https://doi.org/10.1152/jappl.2000.88.1.359.

21. Robert A. Weiss et al., "Clinical Experience with Light-Emitting Diode (LED) Photomodulation," *Dermatologic Surgery* 31, no. 9, pt. 2 (September 2005): 1199–205, https://www.ncbi.nlm.nih.gov/pubmed/16176771.

22. Robert A. Weiss et al., "Clinical Trial of a Novel Non-Thermal LED Array for Reversal of Photoaging: Clinical, Histologic, and Surface Profilometric

Results," *Lasers in Surgery and Medicine* 36, no. 2 (February 2005): 85–91, https://doi.org/10.1002/lsm.20107.

23. Tina S. Alster and Rungsima Wanitphakdeedecha, "Improvement of Post-fractional Laser Erythema with Light-Emitting Diode Photomodulation," *Dermatologic Surgery* 35, no. 5 (May 2009): 813–15, https://doi.org/10.1111 /j.1524-4725.2009.01137.x.

24. M. Maitland DeLand et al., "Treatment of Radiation-Induced Dermatitis with Light-Emitting Diode (LED) Photomodulation," *Lasers in Surgery and Medicine* 39, no. 2 (February 2007): 164–68, https://doi.org/10.1002 /lsm.20455.

25. Disclosure: I founded TrueLight, so I may be biased, and the studies referenced above used different equipment, so they do not apply to TrueLight.

26. Sirous Momenzadeh et al., "The Intravenous Laser Blood Irradiation in Chronic Pain and Fibromyalgia," *Journal of Lasers in Medical Sciences* 6, no. 1 (2015): 6–9, https://doi.org/10.22037/2010.v6i1.7800.

27. Vladimir A. Mikhaylov, "The Use of Intravenous Laser Blood Irradiation (ILBI) at 630–640 nm to Prevent Vascular Diseases and to Increase Life Expectancy," *Laser Therapy* 24, no. 1 (March 31, 2015): 15–26, https://doi .org/10.5978/islsm.15-OR-02.

CHAPTER 6: TURN YOUR BRAIN BACK ON

1. Sue McGreevey, "Brain Checkpoint," Harvard Medical School News and Research, October 25, 2018, https://hms.harvard.edu/news/brain-checkpoint.

2. Brian Giunta et al., "Inflammaging as a Prodrome to Alzheimer's Disease," *Journal of Neuroinflammation* 5 (2008): 51, https://doi.org/10.1186/1742 -2094-5-51.

3. Paul A. Lapchak, "Transcranial Near-Infrared Laser Therapy Applied to Promote Clinical Recovery in Acute and Chronic Neurodegenerative Diseases," *Expert Review of Medical Devices* 9, no. 1 (January 2012): 71–83, https://doi.org/10.1586/erd.11.64.

4. Margaret T. T. Wong-Riley et al., "Photobiomodulation Directly Benefits Primary Neurons Functionally Inactivated by Toxins," *Journal of Biological Chemistry* 280, no. 6 (February 11, 2005): 4761–71, https://doi.org/10.1074 /jbc.M409650200.

5. Javad T. Hashmi et al., "Role of Low-Level Laser Therapy in Neurorehabilitation," *PM&R* 2, no. 12, Supplement 2 (December 2010): S292–S305, https://doi.org/10.1016/j.pmrj.2010.10.013.

6. Michael R. Hamblin, "Shining Light on the Head: Photobiomodulation for Brain Disorders," *BBA Clinical* 6 (October 1, 2016): 113–24, https://doi .org/10.1016/j.bbacli.2016.09.002.

7. Anne Trafton, "Unique Visual Stimulation May Be New Treatment for Alzheimer's," MIT News, December 7, 2016, http://news.mit.edu/2016/visual -stimulation-treatment-alzheimer-1207.

8. Anita E. Saltmarche et al., "Significant Improvement in Cognition in Mild to Moderately Severe Dementia Cases Treated with Transcranial Plus Intranasal Photobiomodulation: Case Series Report," *Journal of Photomedicine and Laser Surgery* 35, no. 8 (August 2017): 432–41, https://doi.org/10.1089/pho.2016.4227.

9. Roger J. Mullins et al., "Insulin Resistance as a Link Between Amyloid-Beta and Tau Pathologies in Alzheimer's Disease," *Frontiers in Aging Neuroscience* 9 (May 3, 2017): 118, https://doi.org/10.3389/fnagi.2017.00118.

10. Patrick Poucheret et al., "Vanadium and Diabetes," *Molecular and Cellular Biochemistry* 188, no. 1–2 (November 1998): 73–80, https://doi.org/10.1023/A:1006820522587.

11. Henry C. Lukaski, "Lessons from Micronutrient Studies in Patients with Glucose Intolerance and Diabetes Mellitus: Chromium and Vanadium," U.S. Department of Health and Human Services, November 8, 2000, https://ods.od.nih.gov/pubs/conferences/lukaski_abstract.html.

12. Kimberly P. Kinzig, Mary Ann Honors, and Sara L. Hargrave, "Insulin Sensitivity and Glucose Tolerance Are Altered by Maintenance on a Ketogenic Diet," *Endocrinology* 151, no. 7 (July 2010): 3105–14, https://doi.org/10.1210/en.2010-0175.

13. John C. Newman and Eric Verdin, "Ketone Bodies as Signaling Metabolites," *Trends in Endocrinology & Metabolism* 25, no. 1 (January 2014): 42–52, https://doi.org/10.1016/j.tem.2013.09.002.

14. Suzanne Craft et al., "Intranasal Insulin Therapy for Alzheimer Disease and Amnestic Mild Cognitive Impairment: A Pilot Clinical Trial," *Archives of Neurology* 69, no. 1 (January 2012): 29–38, https://doi.org/10.1001/archneurol.2011.233.

15. Jill K. Morris and Jeffrey M. Burns, "Insulin: An Emerging Treatment for Alzheimer's Disease Dementia?," *Current Neurology and Neuroscience Reports* 12, no. 5 (October 2012): 520–27, https://doi.org/10.1007/s11910-012-0297-0.

16. Uta Keil et al., "Piracetam Improves Mitochondrial Dysfunction Following Oxidative Stress," *British Journal of Pharmacology* 147, no. 2 (January 2006): 199–208, https://doi.org/10.1038/sj.bjp.0706459.

17. Shelley J. Allen, Judy J. Watson, and David Dawbarn, "The Neurotrophins and Their Role in Alzheimer's Disease," *Current Neuropharmacology* 9, no. 4 (December 2011): 559–73, https://doi.org/10.2174/157015911798376190.

18. Isao Ito et al., "Allosteric Potentiation of Quisqualate Receptors by a Nootropic Drug Aniracetam," *Journal of Physiology* 424 (May 1990): 533–43, https://doi.org/10.1113/jphysiol.1990.sp018081.

19. Richard J. Knapp et al., "Antidepressant Activity of Memory-Enhancing Drugs in the Reduction of Submissive Behavior Model," *European Journal of Pharmacology* 440, no. 1 (April 5, 2002): 27–35, https://doi.org/10.1016/S0014-2999(02)01338-9.

20. Alu Savchenko, N. S. Zakharova, and I. N. Stepanov, "[The Phenotropil Treatment of the Consquences of Brain Organic Lesions]," [Article in Russian] *Zh Nevrol Psikhiatr Im S S Korsakova* 105, no. 12 (2005): 22–26, https://www.ncbi.nlm.nih.gov/pubmed/16447562.

21. "Modfinil," Drugs and Me, http://www.ox.ac.uk/news/2015–08–20-review -%E2%80%98smart-drug%E2%80%99-shows-modafinil-does-enhance -cognition.

22. Paul Newhouse et al., "Intravenous Nicotine in Alzheimer's Disease: A Pilot Study," *Psychopharmacology (Berlin)* 95, no. 2 (1988): 171–75, https://doi .org/10.1007/BF00174504.

23. Paul Newhouse et al., "Nicotine Treatment of Mild Cognitive Impairment: A 6-Month Double-Blind Pilot Clinical Trial," *Neurology* 78, no. 2 (January 10, 2012): 91–101, https://doi.org/10.1212/WNL.0b013e31823efcbb.

24. W. Linert et al., "In Vitro and In Vivo Studies Investigating Possible Antioxidant Actions of Nicotine: Relevance to Parkinson's and Alzheimer's Diseases," *Biochimica et Biophysica Acta* 1454, no. 2 (July 7, 1999): 143–52, https://doi.org/10.1016/S0925-4439(99)00029-0.

25. Toshiharu Nagatsu and Makoto Sawada, "Molecular Mechanism of the Relation of Monoamine Oxidase B and Its Inhibitors to Parkinson's Disease: Possible Implications of Glial Cells," *Journal of Neural Transmission. Supplementum* 71 (2006): 53–65, https://www.ncbi.nlm.nih.gov /pubmed/17447416; Cristina Missale et al., "Dopamine Receptors: From Structure to Function," *Physiological Reviews* 78, no. 1 (January 1998): 189–225, https://doi.org/10.1152/physrev.1998.78.1.189.

26. Claudia Binda et al., "Crystal Structures of Monoamine Oxidase B in Complex with Four Inhibitors of the N-Propargylaminoindan Class," *Journal of Medicinal Chemistry* 47, no. 7 (2004): 1767–74, https://doi.org/10.1021 /jm031087c.

27. M. Jyothi Kumar and Julie K. Andersen, "Perspectives on MAO-B in Aging and Neurological Disease: Where Do We Go from Here?," *Molecular Neurobiology* 30, no. 1 (August 2004): 77–89, https://doi.org/10.1385 /MN:30:1:077; Josep Saura et al., "Biphasic and Region-Specific MAO-B Response to Aging in Normal Human Brain," *Neurobiology of Aging* 18, no. 5 (September–October 1997): 497–507, https://www.ncbi.nlm.nih.gov /pubmed/9390776.

28. E. H. Heinonen and R. Lammintausta, "A Review of the Pharmacology of Selegiline," *Acta Neurologica Scandinavica. Supplementum* 136 (1991): 44–59, https://doi.org/10.1111/j.1600-0404.1991.tb05020.x.

29. Leslie Citrome, Joseph F. Goldberg, and Kimberly Blanchard Portland, "Placing Transdermal Selegiline for Major Depressive Disorder into Clinical Context: Number Needed to Treat, Number Needed to Harm, and Likelihood to Be Helped or Harmed," *Journal of Affective Disorders* 151, no. 2 (November 2013): 409–17, https://doi.org/10.1016/j.jad.2013.06.027.

30. Carolina M. Maier and Pak H. Chan, "Role of Superoxide Dismutases in Oxidative Damage and Neurodegenerative Disorders," *Neuroscientist* 8, no. 4 (August 2002): 323–34, https://doi.org/10.1177/107385840200800408.

31. Norton W. Milgram et al., "Maintenance on L-Deprenyl Prolongs Life in Aged Male Rats," *Life Sciences* 47, no. 5 (1990): 415–20, https://doi .org/10.1016/0024-3205(90)90299-7; Kenichi Kitani et al., "(-)Deprenyl Increases the Life Span as Well as Activities of Superoxide Dismutase and Catalase but Not of Glutathione Peroxidase in Selective Brain Regions in Fischer Rats," *Annals of the New York Academy of Sciences* 717 (June 30, 1994): 60–71, https://doi.org/10.1111/j.1749-6632.1994.tb12073.x.

32. Joseph Knoll, "The Striatal Dopamine Dependency of Life Span in Male Rats. Longevity Study with (-)Deprenyl," *Mechanisms of Ageing and Development* 46, no. 1–3 (December 1988): 237–62, https://doi.org/10.1016/0047 -6374(88)90128-5.

33. Joseph Knoll, "The Striatal Dopamine Dependency."

34. Giovanni Ghirlanda et al., "Evidence of Plasma CoQ10-Lowering Effect by HMG-CoA Reductase Inhibitors: A Double-Blind, Placebo-Controlled Study," *Journal of Clinical Pharmacology* 33, no. 3 (1993): 226–29, https ://doi.org/10.1002/j.1552-4604.1993.tb03948.x.

35. Sausan Jaber and Brian M. Polster, "Idebenone and Neuroprotection: Antioxidant, Pro-Oxidant, or Electron Carrier?," *Journal of Bioenergetics and Biomembranes* 47, no. 1–2 (2014): 111–8, https://doi.org/10.1007/s10863-014-9571-y.

36. X. J. Liu and W. T. Wu, "Effects of Ligustrazine, Tanshinone II A, Ubiquinone, and Idebenone on Mouse Water Maze Performance," *Zhongguo Yao Li Xue Bao* 20, no. 11 (November 1999): 987–90, https://www.ncbi.nlm .nih.gov/pubmed/11270979.

37. K. Murase et al., "Stimulation of Nerve Growth Factor Synthesis/Secretion in Mouse Astroglial Cells by Coenzymes," *Biochemistry and Molecular Biology International* 30, no. 4 (July 1993): 615–21, https://www.ncbi.nlm .nih.gov/pubmed/8401318.

38. Natsumi Noji et al., "Simple and Sensitive Method for Pyrroloquinoline Quinone (PQQ) Analysis in Various Foods Using Liquid Chromatography/ Electrospray-Ionization Tandem Mass Spectrometry," *Journal of Agricultural and Food Chemistry* 55, no. 18 (September 5, 2007): 7258–63, https ://doi.org/10.1021/jf070483r.

39. K. A. Bauerly et al., "Pyrroloquinoline Quinone Nutritional Status Alters Lysine Metabolism and Modulates Mitochondrial DNA Content in the Mouse and Rat," *Biochimica et Biophysica Acta* 1760, no. 11 (November 2006): 1741–48, https://doi.org/10.1016/j.bbagen.2006.07.009.

40. Calliandra B. Harris et al., "Dietary Pyrroloquinoline Quinone (PQQ) Alters Indicators of Inflammation and Mitochondrial-Related Metabolism in Human Subjects," *The Journal of Nutritional Biochemistry* 24, no. 12 (December 2013): 2076–84, https://doi.org/10.1016/j.jnutbio.2013.07.008.

41. K. Bauerly et al., "Altering Pyrroloquinoline Quinone Nutritional Status Modulates Mitochondrial, Lipid, and Energy Metabolism in Rats," *PLoS One* 6, no. 7 (2011): e21779, https://doi.org/10.1371/journal.pone.0021779.

42. Kana Nunome et al., "Pyrroloquinoline Quinone Prevents Oxidative Stress-Induced Neuronal Death Probably Through Changes in Oxidative Status of DJ-1," *Biological and Pharmaceutical Bulletin* 31, no. 7 (July 2008): 1321–26, https://doi.org/10.1248/bpb.31.1321.

43. Francene M. Steinberg, M. Eric Gershwin, and Robert B. Rucker, "Dietary Pyrroloquinoline Quinone: Growth and Immune Response in BALB/c Mice," *The Journal of Nutrition* 124, no. 5 (May 1994): 744–53, https://doi.org/10.1093/jn/124.5.744.

44. Kei Ohwada et al., "Pyrroloquinoline Quinone (PQQ) Prevents Cognitive Deficit Caused by Oxidative Stress in Rats," *Journal of Clinical Biochemistry and Nutrition* 42, no. 1 (January 2008): 29–34, https://doi.org/10.3164/jcbn.2008005.

45. Bo-qing Zhu et al., "Pyrroloquinoline Quinone (PQQ) Decreases Myocardial Infarct Size and Improves Cardiac Function in Rat Models of Ischemia and Ischemia/Reperfusion," *Cardiovascular Drugs and Therapy* 18, no. 6 (November 2004): 421–31, https://doi.org/10.1007/s10557-004-6219-x.

46. Pere Puigserver, "Tissue-Specific Regulation of Metabolic Pathways Through the Transcriptional Coactivator PGC1-alpha," *International Journal of Obesity* 29, Supplement 1 (March 2005): S5–S9, https://doi.org/10.1038/sj.ijo.0802905.

47. Chanoch Miodownik et al., "Serum Levels of Brain-Derived Neurotrophic Factor and Cortisol to Sufate of Dehydroepiandrosterone Molar Ratio Associated with Clinical Response to L-Theanine as Augumentation of Antipsychotic Therapy in Schizophrenia and Schizoaffective Disorder Patients," *Clinical Neuropharmacology* 34, no. 4 (July–August 2011): 155–60, https://doi.org/10.1097/WNF.0b013e318220d8c6.

48. Kenta Kimura et al., "L-Theanine Reduces Psychological and Physiological Stress Responses," *Biological Psychology* 74, no. 1 (January 2007): 39–45, https://doi.org/10.1016/j.biopsycho.2006.06.006.

49. Anna Christina Nobre, Anling Rao, and Gail N. Owen, "L-Theanine, a Natural Constituent in Tea, and Its Effect on Mental State," *Asia Pacific Journal of Clinical Nutrition* 17, Supplement 1 (2008): 167–68, https://www.ncbi.nlm.nih.gov/pubmed/18296328.

50. Crystal F. Haskell et al., "The Effects of L-Theanine, Caffeine and Their Combination on Cognition and Mood," *Biological Psychology* 77, no. 2 (February 2008): 113–22, https://doi.org/10.1016/j.biopsycho.2007.09.008.

51. Puei-Lene Lai et al., "Neurotrophic Properties of the Lion's Mane Medicinal Mushroom, *Hericium erinaceus* (Higher Basidiomycetes) from Malaysia," *International Journal of Medicinal Mushrooms* 15, no. 6 (2013): 539–54, https://doi.org/10.1615/IntJMedMushr.v15.i6.30.

52. Leigh Hopper, "Curcumin Improves Memory and Mood, New UCLA Study Says," UCLA Newsroom, January 22, 2018, http://newsroom.ucla .edu/releases/curcumin-improves-memory-and-mood-new-ucla-study -says.

53. Annu Khajuria, N. Thusu, and U. Zutshi, "Piperine Modulates Permea- bility Characteristics of Intestine by Inducing Alterations in Membrane Dynamics: Influence on Brush Border Membrane Fluidity, Ultrastructure and Enzyme Kinetics," *Pytomedicine* 9, no. 3 (April 2002): 224–31, https ://doi.org/10.1078/0944-7113-00114.

54. Guy-Armel Bounda and Yu Feng, "Review of Clinical Studies of *Polygo- num multiflorum* Thunb. and Its Isolated Bioactive Compounds," *Phar- macognosy Research* 7, no. 3 (July–September 2015): 225–36, https://doi .org/10.4103/0974-8490.157957.

55. Hye Jin Park, Nannan Zhang, and Dong Ki Park, "Topical Application of *Polygonum multiflorum* Extract Induces Hair Growth of Resting Hair Fol- licles Through Upregulating Shh and β-Catenin Expression in C57BL/6 Mice," *Journal of Ethnopharmacology* 135, no. 2 (May 17, 2011): 369–75, https://doi.org/10.1016/j.jep.2011.03.028; Ya Nan Sun et al., "Promotion Effect of Constituents from the Root of *Polygonum multiflorum* on Hair Growth," *Bioorganic & Medicinal Chemistry Letters* 23, no. 17 (September 1, 2013): 4801–05, https://doi.org/10.1016/j.bmcl.2013.06.098.

CHAPTER 7: METAL BASHING

1. Tchounwou et al., "Heavy Metal."

2. Monisha Jaishankar et al., "Toxicity, Mechanism and Health Effects of Some Heavy Metals," *Interdisciplinary Toxicology* 7, no. 2 (June 2014): 60– 72, https://doi.org/10.2478/intox-2014-0009.

3. "Lead Poisoning and Health," World Health Organization (WHO), August 23, 2018, http://www.who.int/news-room/fact-sheets/detail/lead-poisoning -and-health.

4. Bruce P. Lanphear et al., "Low-Level Lead Exposure and Mortality in US Adults: A Population-Based Cohort Study," *The Lancet: Public Health* 3, no. 4 (April 1, 2018): PE177–E184, https://doi.org/10.1016/S2468 -2667(18)30025-2.

5. Petra Cvjetko, Ivan Cvjetko, and Mirjana Pavlica, "Thallium Toxicity in Humans," *Arh Hig Rada Toksikol* 61, no. 1 (March 2010): 111–19, https ://doi.org/10.2478/10004-1254-61-2010-1976.

6. J. Pavlíčková et al., "Uptake of Thallium from Artificially Contaminated Soils by Kale (*Brassica oleracea* L. var. acephala)," *Plant, Soil and Environ- ment* 52, no. 12 (December 2006): 484–91, https://doi.org/10.17221/3545 -PSE.

7. Yanlong Jia et al., "Thallium at the Interface of Soil and Green Cabbage (*Brassica oleracea* L. var. capitata L.): Soil-Plant Transfer and Influencing

Factors," *Science of the Total Environment* 450–51 (April 15, 2013): 140–47, https://doi.org/10.1016/j.scitotenv.2013.02.008.

8. Zenping Ning et al., "High Accumulation and Subcellular Distribution of Thallium in Green Cabbage (*Brassica oleracea* L. Var. Capitata L.)," *International Journal of Phytoremediation* 17, no. 11 (2015): 1097–104, https://doi.org/10.1080/15226514.2015.1045133.

9. Sung Kyun Park et al., "Associations of Blood and Urinary Mercury with Hypertension in U.S. Adults: The NHANES 2003–2006," *Environmental Research* 123 (May 2013): 25–32, https://doi.org/10.1016/j.envres.2013.02.003; Mark C. Houston, "Role of Mercury Toxicity in Hypertension, Cardiovascular Disease, and Stroke," *Journal of Clinical Hypertension* 13, no. 8 (August 2011): 621–27, https://doi.org/10.1111/j.1751-7176.2011.00489.x.

10. Arif Tasleem Jan et al., "Heavy Metals and Human Health: Mechanistic Insight into Toxicity and Counter Defense System of Antioxidants," *International Journal of Molecular Sciences* 16, no. 12 (2015): 29592–630, https://doi.org/10.3390/ijms161226183.

11. Margaret E. Sears, "Chelation: Harnessing and Enhancing Heavy Metal Detoxification—A Review," *The Scientific World Journal* 2013 (March 14, 2013): 219840, https://doi.org/10.1155/2013/219840.

12. Sears, "Chelation,"; Alan Becker and Karam F. A. Soliman, "The Role of Intracellular Glutathione in Inorganic Mercury-Induced Toxicity in Neuroblastoma Cells," *Neurochemical Research* 34, no. 9 (September 2009): 1677–84, https://doi.org/10.1007/s11064-009-9962-3.

13. Lambros Kromidas, Louis David Trombetta, and Ijaz Siraj Jamall, "The Protective Effects of Glutathione Against Methymercury Cytotoxicity," *Toxicology Letters* 51, no. 1 (March 1990): 67–80, https://doi.org/10.1016/0378-4274(90)90226-C.

14. Ralf Dringen, "Metabolism and Functions of Glutathione in Brain," *Progress in Neurobiology* 62, no. 6 (December 2000): 649–71, https://doi.org/10.1016/S0301-0082(99)00060-X.

15. Danyelle M. Townsend, Kenneth D. Tew, and Haim Tapiero, "The Importance of Glutathione in Human Disease," *Biomedicine & Pharmacotherapy* 57, no. 3–4 (May–June 2003): 145–55, https://doi.org/10.1016/S0753-3322(03)00043-X.

16. Lester Packer, Hans J. Tritschler, and Klaus Wessel, "Neuroprotection by the Metabolic Antioxidant Alpha-Lipoic Acid," *Free Radical Biology and Medicine* 22, no. 1–2 (1997): 359–78, https://doi.org/10.1016/s0891-5849(96)00269-9.

17. Packer, Tritschler, and Wessel, "Neuroprotection."

18. D. Ziegler et al., "Treatment of Symptomatic Diabetic Polyneuropathy with the Antioxidant Alpha-Lipoic Acid: A 7-Month Multicenter Randomized Controlled Trial (ALADIN III Study). ALADIN III Study Group. Alpha-Lipoic Acid in Diabetic Neuropathy," *Diabetes Care* 22, no. 8 (August 1999): 1296–301, https://doi.org/10.1111/j.1464-5491.2004.01109.x.

19. Mostafa I. Waly, Zahir Humaid Al Attabi, and Nejib Guizani, "Low Nourishment of Vitamin C Induces Glutathione Depletion and Oxidative Stress in Healthy Young Adults," *Preventive Nutrition and Food Science* 20, no. 3 (September 2015): 198–203, https://doi.org/10.3746/pnf.2015.20.3.198.

20. C. S. Johnston, C. G. Meyer, and J. C. Srilakshmi, "Vitamin C Elevates Red Blood Cell Glutathione in Healthy Adults," *American Journal of Clinical Nutrition* 58, no. 1 (July 1993): 103–105, https://doi.org/10.1093/ajcn/58.1.103.

21. Hoseob Lihm et al., "Vitamin C Modulates Lead Excretion in Rats," *Anatomy & Cell Biology* 46, no. 4 (2013): 239–45, https://doi.org/10.5115/acb.2013.46.4.239.

22. Tina Tinkara Peternelj and Jeff S. Coombes, "Antioxidant Supplementation During Exercise Training: Beneficial or Detrimental?," *Sports Medicine* 41, no. 12 (December 1, 2011): 1043–69, https://doi.org/10.2165/11594400-000000000-00000.

23. Christy C. Bridges and Rudolfs K. Zalups, "Molecular and Ionic Mimicry and the Transport of Toxic Metals," *Toxicology and Applied Pharmacology* 204, no. 3 (May 2005): 274–308, https://doi.org/10.1201/9781420059984-c10.

24. V. V. Frolkis et al., "Effect of Enterosorption on Animal Lifespan," *Biomaterials, Artificial Cells and Artificial Organs* 17, no. 3 (1989): 341–51, https://doi.org/10.3109/10731198909118290.

25. Pasi Kuusisto et al., "Effect of Activated Charcoal on Hypercholesterolaemia," *The Lancet* 2, no. 8503 (August 16, 1986): 366–67, https://doi.org/10.1016/S0140-6736(86)90054-1.

26. "Activated Carbon: An Overview," ScienceDirect, https://www.sciencedirect.com/topics/pharmacology-toxicology-and-pharmaceutical-science/activated-carbon.

27. Antonello Santini and Alberto Ritieni, "Aflatoxins: Risk, Exposure and Remediation," in *Aflatoxins—Recent Advances and Future Prospects*, ed. Mehdi Razzaghi-Abyaneh (IntechOpen, January 23, 2013), https://www.intechopen.com/books/aflatoxins-recent-advances-and-future-prospects/aflatoxins-risk-exposure-and-remediation.

28. Takuya Uchikawa et al., "Enhanced Elimination of Tissue Methymercury in *Parachlorella beijerinckii*-Fed Mice," *Journal of Toxicological Sciences* 36, no. 1 (January 2011): 121–26, https://doi.org/10.2131/jts.36.121.

29. Dorothy A. Kieffer, Roy J. Martin, and Sean H. Adams, "Impact of Dietary Fibers on Nutrient Management and Detoxification Organs: Gut, Liver, and Kidneys," *Advances in Nutrition* 7, no. 6 (November 2016): 1111–21, https://doi.org/10.3945/an.116.013219.

30. Isaac Eliaz et al., "The Effect of Modified Citrus Pectin on Urinary Excretion of Toxic Elements," *Phytotherapy Research* 20, no. 10 (October 2006): 849–64, https://doi.org/10.1002/ptr.1953.

31. Vladislav V. Glinsky and Avraham Raz, "Modified Citrus Pectin Anti-Metastatic Properties: One Bullet, Multiple Targets," *Carbohydrate Research* 344, no. 14 (September 28, 2008): 1788–91, https://doi.org/10.1016/j.carres.2008.08.038.

32. Steven De Berg, "A Lifesaving Nutrient in Citrus Fruit," LifeExtension, October 2014, https://www.lifeextension.com/magazine/2014/10/why-some-people-need-modified-citrus-pectin/page-01.

33. Lu-Gang Yu et al., "Galectin-3 Interaction with Thomsen-Friedenreich Disaccharide on Cancer-Associated MUC1 Causes Increased Cancer Cell Endothelial Adhesion," *Journal of Biological Chemistry* 282, no. 1 (January 5, 2007): 773–81, https://doi.org/10.1074/jbc.M606862200; Qicheng Zhao et al., "Circulating Galectin-3 Promotes Metastasis by Modifying MUC1 Localization on Cancer Cell Surface." *Cancer Research* 69, no. 17 (September 1, 2009): 6799–806, https://doi.org/10.1158/0008-5472.CAN-09-1096; Maria Kolatsi-Joannou et al., "Modified Citrus Pectin Reduces Galectin-3 Expression and Disease Severity in Experimental Acute Kidney Injury," *PLoS One* 6, no. 4 (2011): e18683, https://doi.org/10.1371/journal.pone.0018683; Dirk J. A. Lok et al., "Prognostic Value of Galectin-3, a Novel Marker of Fibrosis, in Patients with Chronic Heart Failure: Data from the DEAL-HF Study," *Clinical Research in Cardiology* 99, no. 5 (May 2010): 323–28, https://doi.org/10.1007/s00392-010-0125-y.

34. Gervasio A. Lamas et al., "Heavy Metals, Cardiovascular Disease, and the Unexpected Benefits of Chelation Therapy," *Journal of the American College of Cardiology* 67, no. 20 (May 24, 2016): 2411–18, https://doi.org/10.1016/j.jacc.2016.02.066.

35. Margaret E. Sears, Kathleen J. Kerr, and Riina I. Bray, "Arsenic, Cadmium, Lead, and Mercury in Sweat: A Systematic Review," *Journal of Environmental and Public Health* 2012 (2012): 184745, https://doi.org/10.1155/2012/184745.

36. Larry A. Tucker, "Physical Activity and Telomere Length in U.S. Men and Women: An NHANES Investigation," *Preventive Medicine* 100 (July 2017): 145–51, https://doi.org/10.1016/j.ypmed.2017.04.027.

CHAPTER 8: POLLUTING YOUR BODY WITH OZONE

1. Renate Viebahn-Haensler, *The Use of Ozone in Medicine*, 4th ed. (Medicina Biologica, 2002).

2. Zullyt Zamora Rodríguez et al., "Preconditioning with Ozone/Oxygen Mixture Induces Reversion of Some Indicators of Oxidative Stress and Prevents Organic Damage in Rats with Fecal Peritonitis," *Inflammation Research* 58, no. 7 (July 2009): 371–75, https://doi.org/10.1007/s00011-009-0001-2.

3. Robert J. Rowen, "Ozone Therapy as a Primary and Sole Treatment for Acute Bacterial Infection: Case," *Medical Gas Research* 8, no. 3 (July–September 2018): 121–24, https://doi.org/10.4103/2045-9912.241078.

4. Robert J. Rowen et al., "Rapid Resolution of Hemorrhagic Fever (Ebola)

in Sierra Leone with Ozone Therapy," *African Journal of Infectious Diseases (AJID)* 10, no. 1 (August 1, 2015): 45–59, https://doi.org/10.21010/ajid .v10i1.10.

5. Michael B. Schultz and David A. Sinclair, "Why NAD(+) Declines During Aging: It's Destroyed," *Cell Metabolism* 23, no. 6 (June 14, 2016): 965–66, https://doi.org/10.1016/j.cmet.2016.05.022.

6. Christian T. Sheline, M. Margarita Behrens, and Dennis W. Choi, "Zinc-Induced Cortical Neuronal Death: Contribution of Energy Failure Attributable to Loss of NAD+ and Inhibition of Glycolysis," *Journal of Neuroscience* 20, no. 9 (May 1, 2000): 3139–46, https://doi.org/10.1523/JNEU ROSCI.20-09-03139.2000.

7. Leonard Guarente, "Sirtuins in Aging and Disease," *Cold Spring Harbor Symposia on Quantitative Biology* 72 (2007): 483–88, https://doi.org /10.1101/sqb.2007.72.024.

8. Eriko Michishita et al., "SIRT6 Is a Histone H3 Lysine 9 Deacetylase That Modulates Teomeric Chromatin," *Nature* 452, no. 7186 (March 27, 2008): 492–96, https://doi.org/10.1038/nature06736.

9. Hongying Yang et al., "Nutrient-Sensitive Mitochondrial NAD+ Levels Dictate Cell Survival," *Cell* 130, no. 6 (September 21, 2007): 1095–107, https://doi.org/10.1016/j.cell.2007.07.035.

10. Suping Wang et al., "Cellular NAD Replenishment Confers Marked Neuroprotection Against Ischemic Cell Death: Role of Enhanced DNA Repair," *Stroke* 39, no. 9 (September 2008): 2587–95, https://doi.org/10.1161 /STROKEAHA.107.509158.

11. Sydney Shall, "ADP-Ribose in DNA Repair: A New Component of DNA Excision Repair," *Advances in Radiation Biology* 11 (1984): 1–69 https://doi .org/10.1016/B978-0-12-035411-5.50007-1.

12. Shall, "ADP-Ribose."

13. Evandro Fei Fang et al., "NAD Replenishment Improves Lifespan and Healthspan in Ataxia Telangiectasia Models via Mitophagy and DNA Repair," *Cell Metabolism* 24, no. 4 (October 11, 2016): 578, fig. 7, https://doi .org/10.1016/j.cmet.2016.09.004.

14. Hassina Massudi et al., "Age-Associated Changes in Oxidative Stress and NAD Metabolism in Human Tissue," *PLoS One* 7, no. 7 (July 2012): e42357, fig. 4, https://doi.org/10.1371/journal.pone.0042357.

15. Massudi et al., "Age-Associated Changes," e42357.

16. Jun Yoshino et al., "Nicotinamide Mononucleotide, a Key NAD(+) Intermediate, Treats the Pathophysiology of Diet- and Age-Induced Diabetes in Mice," *Cell Metabolism* 14, no. 4 (October 5, 2011): 528–36, https://doi .org/10.1016/j.cmet.2011.08.014.

17. Péter Bai et al., "PARP-1 Inhibition Increases Mitochondrial Metabolism Through SIRT1 Activation," *Cell Metabolism* 13, no. 4 (April 6, 2011): 461–68, https://doi.org/10.1016/j.cmet.2011.03.004.

18. Hongbo Zhang et al., "NAD Repletion Improves Mitochondrial and Stem Cell Function and Enhances Life Span," *Science* 352, no. 6292 (June 17, 2016): 1436–43, https://doi.org/10.1126/science.aaf2693.

19. Satoru Hayashida et al., "Fasting Promotes the Expression of SIRT1, an NAD+-Dependent Protein deacetylase, via Activation of PPARα in Mice," *Molecular and Cellular Biochemistry* 339, no. 1–2 (June 2010): 285–92, https://doi.org/10.1007/s11010-010-0391-z.

20. David S. Williams et al., "Oxalocetate Supplementation Increases Lifespan in *Caenorhabditis elegans* Through an AMPK/FOXO-Dependent Pathway," *Aging Cell* 8, no. 6 (December 2009): 765–68, https://doi.org/10.1111/j.1474-9726.2009.00527.x.

CHAPTER 9: FERTILITY = LONGEVITY

1. C. C. Zouboulis and E. Makrantonaki, "Hormonal Therapy of Intrinsic Aging," *Rejuvenation Research* 15, no. 3 (June 2012): 302–12, https://doi.org/10.1089/rej.2011.1249.

2. Cynthia K. Sites, "Bioidentical Hormones for Menopausal Therapy," *Women's Health* 4, no. 2 (March 2008): 163–71, https://doi.org/10.2217/17455057.4.2.163.

3. Peter J. Snyder et al., "Effect of Testosterone Treatment on Body Composition and Muscle Strength in Men Over 65 Years of Age," *Journal of Clinical Endocrinology & Metabolism* 84, no. 8 (August 1, 1999): 2647–53, https://doi.org/10.1210/jcem.84.8.5885.

4. Anne M. Kenny et al., "Effects of Transdermal Testosterone on Cognitive Function and Health Perception in Older Men with Low Bioavailable Testosterone Levels," *Journals of Gerontology. Series A, Biological Sciences and Medical Sciences* 57, no. 5 (May 2002): M321–25, https://doi.org/10.1093/gerona/57.5.M321.

5. Giuseppe M. Rosano et al., "Low Testosterone Levels Are Associated with Coronary Artery Disease in Male Patients with Angina," *International Journal of Impotence Research* 19, no. 2 (March–April 2007): 176–82, https://doi.org/10.1038/sj.ijir.3901504.

6. Rishi Sharma et al., "Normalization of Testosterone Level Is Associated with Reduced Incidence of Myocardial Infarction and Mortality In Men," *European Heart Journal* 36, no. 40 (October 21, 2015): 2706–15, https://doi.org/10.1093/eurheartj/ehv346.

7. Nikolaos Samaras et al., "Off-Label Use of Hormones as an Antiaging Strategy: A Review," *Clinical Interventions in Aging* 9 (July 23, 2014): 1175–86, https://doi.org/10.2147/CIA.S48918.

8. Jacques E. Rossouw et al., "Risks and Benefits of Estrogen Plus Progestin in Healthy Postmenopausal Women: Principal Results from the Women's Health Initiative Randomized Controlled Trial," *JAMA* 288, no. 3 (July 17, 2002): 321–33, https://doi.org/10.1001/jama.288.3.321.

9. Samaras et al., "Off-Label Use."

10. Michael Castleman, "The Prescription for a Longer Life? More Sex," *Psychology Today*, May 15, 2017, https://www.psychologytoday.com/ca/blog/all-about-sex/201705/the-prescription-longer-life-more-sex.

11. Samuel S. C. Yen, "Dehydroepiandrosterone Sulfate and Longevity: New Clues for an Old Friend," *Proceedings of the National Academy of Sciences of the USA* 98, no. 15 (2001): 8167–69, https://doi.org/10.1073/pnas.161278698.

12. Alessandro D. Genazzani, Chiara Lanzoni, and Andrea R. Genazzani, "Might DHEA Be Considered a Beneficial Replacement Therapy in the Elderly?," *Drugs & Aging* 24, no. 3 (2007): 173–85, https://doi.org/10.2165/00002512-200724030-00001.

13. M. Murad Basar et al., "Relationship Between Serum Sex Steroids and Aging Male Symptoms Score and International Index of Erectile Function," *Urology* 66, no. 3 (September 2005): 597–601, https://doi.org/10.1016/j.urology.2005.03.060.

14. Sun-Ouck Kim et al., "Penile Growth in Response to Human Chorionic Gonadotropin (HCG) Treatment in Patients with Idiopathic Hypogonadotrophic Hypogonadism," *Chonnam Medical Journal* 47, no. 1 (April 2011): 39–42, https://doi.org/10.4068/cmj.2011.47.1.39.

15. Christian Elabd et al., "Oxytocin Is an Age-Specific Circulating Hormone That Is Necessary for Muscle Maintenance and Regeneration," *Nature Communications* 5, (2014): 4082, https://doi.org/10.1038/ncomms5082.

16. Jean-Jacques Legros, "Inhibitory Effect of Oxytocin on Corticotrope Function in Humans: Are Vasopressin and Oxytocin Ying-Yang Neurohormones?," *Psychoneuroendocrinology* 26, no. 7 (2001): 649–55, https://doi.org/10.1016/S0306-4530(01)00018-X.

17. Thomas G. Travison et al., "A Population-Level Decline in Serum Testosterone Levels in American Men," *Journal of Clinical Endocrinology & Metabolism* 92, no. 1 (January 2007): 196–202, https://doi.org/10.1210/jc.2006–1375.

18. Jeff S. Volek et al., "Testosterone and Cortisol in Relationship to Dietary Nutrients and Resistance Exercise," *Journal of Applied Physiology* 82, no. 1 (1997): 49–54, https://doi.org/10.1152/jappl.1997.82.1.49.

19. Esa Hämäläinen et al., "Diet and Serum Sex Hormones in Healthy Men," *Journal of Steroid Biochemistry* 20, no. 1 (1984): 459–64, https://doi.org/10.1016/0022-4731(84)90254-1

20. E. Wehr et al., "Association of Vitamin D Status with Serum Androgen Levels in Men," *Clinical Endocrinology* 73, no. 2 (August 2010): 243–48, https://doi.org/10.1111/j.1365-2265.2009.03777.x.

21. Susan Jobling et al., "A Variety of Environmentally Persistent Chemicals, Including Some Phthalate Plasticizers, Are Weakly Estrogenic," *Environmental Health Perspectives* 103, no. 6 (June 1995): 582–87, https://doi.org/10.1289/ehp.95103582.

22. Edwin J. Routledge et al., "Some Alkyl Hydroxy Benzoate Preservatives (Parabens) Are Estrogenic," *Toxicology and Applied Pharmacology* 153, no. 1 (December 1998): 12–19, https://doi.org/10.1006/taap.1998.8544.

23. Katrina Woznicki, "Birth Control Pills Put Brakes on Women's Sex Drive," WebMD, May 5, 2010, https://www.webmd.com/sex/birth-control/news/20100505/birth-control-pills-put-brakes-on-womens-sex-drive#2.

24. Claudia Panzer et al., "Impact of Oral Contraceptives on Sex Hormone-Binding Globulin and Androgen Levels: A Retrospective Study in Women with Sexual Dysfunction," *Journal of Sexual Medicine* 3, no. 1 (January 2006): 104–13, https://doi.org/10.1111/j.1743-6109.2005.00198.x.

25. William J. Kraemer et al., "Endogenous Anabolic Hormonal and Growth Factor Responses to Heavy Resistance Exercises in Males and Females," *International Journal of Sports Medicine* 12, no. 2 (May 1991): 228–35, https://doi.org/10.1055/s-2007-1024673.

26. Patrick Wahl, "Hormonal and Metabolic Responses to High Intensity Interval Training," *Journal of Sports Medicine & Doping Studies* 3 (January 24, 2013): e132, https://doi.org/10.4172/2161-0673.1000e132.

27. European Society of Cardiology, "Endurance but Not Resistance Training Has Anti-Aging Effects," EurekAlert!, November 27, 2018, https://www.eurekalert.org/pub_releases/2018-11/esoc-ebn112618.php.

28. Salam Ranabir and Reetu Keisam, "Stress and Hormones," *Indian Journal of Endocrinology and Metabolism* 15, no. 1 (2011): 18–22, https://doi.org/10.4103/2230-8210.77573.

29. Andrew B. Dollins et al., "L-Tyrosine Ameliorates Some Effects of Lower Body Negative Pressure Stress," *Physiology & Behavior* 57, no. 2 (February 1995): 223–30, https://doi.org/10.1016/0031-9384(94)00278-D.

30. Yue-Feng Chen and Martin Gerdes, "Deadly Connection: Hypothyroidism and Heart Disease," *Diagnostic and Interventional Cardiology*, March 15, 2007, https://www.dicardiology.com/article/deadly-connection-hypothyroidism-and-heart-disease.

CHAPTER 10: YOUR TEETH ARE A WINDOW TO THE NERVOUS SYSTEM

1. Barbara Gefvert, "Medical Lasers/Neuroscience: Photobiomodulation and the Brain: Traumatic Brain Injury and Beyond," BioOptics World, May 9, 2016, https://www.biopticsworld.com/articles/print/volume-9/issue-5/medical-lasers-neuroscience-photobiomodulation-and-the-brain-traumatic-brain-injury-and-beyond.html.

2. Lydia E. Kuo et al., "Chronic Stress, Combined with a High-Fat/High-Sugar Diet, Shifts Sympathetic Signaling Toward Neuropeptide Y and Leads to Obesity and the Metabolic Syndrome," *Annals of the New York Academy of Sciences* 1148 (December 2008): 232–37, https://doi.org/10.1196/annals.1410.035.

3. Keith N. Frayn, "Visceral Fat and Insulin Resistance—Causative or Cor-

relative?," *British Journal of Nutrition* 83, Supplement 1 (March 2000): S71–77, https://doi.org/10.1017/S0007114500000982.

4. Ken Kishida et al., "Relationships Between Circulating Adiponectin Levels and Fat Distribution in Obese Subjects," *Journal of Atherosclerosis and Thrombosis* 18, no. 7 (2011): 592–95, https://doi.org/10.5551/jat.7625.

5. Yumi Matsushita et al., "Adiponectin and Visceral Fat Associate with Cardiovascular Risk Factors," *Obesity* 21 (2014): 287–91, https://doi.org/10.1002/oby.20425.

6. Jeb S. Orr et al., "Large Artery Stiffening with Weight Gain in Humans: Role of Visceral Fat Accumulation," *Hypertension* 51, no. 6 (June 2008): 1519–24, https://doi.org/10.1161/HYPERTENSIONAHA.108.112946.

7. Christopher K. Kepler et al., "Substance P Stimulates Production of Inflammatory Cytokines in Human Disc Cells," *Spine* 38, no. 21 (October 1, 2013): E1291–99, https://doi.org/10.1097/BRS.0b013e3182a42bc2.

8. Mengmeng Zhan et al., "Upregulated Expression of Substance P (SP) and NK1R in Eczema and SP-Induced Mast Cell Accumulation," *Cell Biology and Toxicology* 33, no. 4 (August 2017): 389–405, https://doi.org/10.1007/s10565-016-9379-0; Beni Amatya et al., "Expression of Tachykinins and Their Receptors in Plaque Psoriasis with Pruritus," *British Journal of Dermatology* 164, no. 5 (May 2011): 1023–29, https://doi.org/10.1111/j.1365-2133.2011.10241.x.

9. Terence M. O'Connor et al., "The Role of Substance P in Inflammatory Disease," *Journal of Cellular Physiology* 201, no. 2 (November 2004): 167–80, https://doi.org/10.1002/jcp.20061.

10. Miguel Muñoz and Rafael Coveñas, "Involvement of Substance P and the NK-1 Receptor in Cancer Progression," *Peptides* 48 (October 2013): 1–9, https://doi.org/10.1016/j.peptides.2013.07.024.

11. Pranela Rameshwar and Pedro Gascón, "Substance P (SP) Mediates Production of Stem Cell Factor and Interleukin-1 in Bone Marrow Stroma: Potential Autoregulatory Role for These Cytokines in SP Receptor Expression and Induction," *Blood* 86, no. 2 (July 1995): 482–90, https://www.ncbi.nlm.nih.gov/pubmed/7541664.

12. Thomas F. Burks, Stephen H. Buck, and Matthew S. Miller, "Mechanisms of Depletion of Substance P by Capsaicin," *Federation Proceedings* 44, no. 9 (1985): 2531–34, https://www.ncbi.nlm.nih.gov/pubmed/2581820.

13. P. Anand and Keith Bley, "Topical Capsaicin for Pain Management: Therapeutic Potential and Mechanisms of Action of the New High-Concentration Capsaicin 8% Patch," *British Journal of Anaesthesia* 107, no. 4 (October 2011): 490–502, https://doi.org/10.1093/bja/aer260.

14. Sharath Asokan et al., "Effect of Oil Pulling on *Streptococcus mutans* Count in Plaque and Saliva Using Dentocult SM Strip Mutans Test: A Randomized, Controlled, Triple-Blind Study," *Journal of Indian Society of Pedodontics and Preventive Dentistry* 26, no. 1 (March 2008): 12–17, https://www.ncbi.nlm.nih.gov/pubmed/18408265.

15. Sharath Asokan, Raghuraman Chamundeswari, and Pamela Emmadi, "Effect of Oil Pulling on Plaque Induced Gingivitis: A Randomized, Controlled, Triple-Blind Study," *Indian Journal of Dental Research* 20, no. 1 (January 2009): 47–51, https://doi.org/10.4103/0970-9290.49067.

16. Asokan, Chamundeswari, and Emmadi, "Effect of Oil Pulling."

17. M. K. Nair et al., "Antibacterial Effect of Caprylic Acid and Monocaprylin on Major Bacterial Mastitis Pathogens," *Journal of Dairy Science* 88, no. 10 (October 2005): 3488–95, https://doi.org/10.3168/jds.S0022-0302(05)73033-2.

18. Foundation of the National Lipid Association, Learn Your Lipids, http://www.learnyourlipids.com/lipids/.

19. Radka Hulankova, Gabriela Borilova, and Iva Steinhauserova, "Combined Antimicrobial Effect of Oregano Essential Oil and Caprylic Acid in Minced Beef," *Meat Science* 95, no. 2 (October 2013): 190–94, https://doi.org/10.1016/j.meatsci.2013.05.003.

CHAPTER II: HUMANS ARE WALKING PETRI DISHES

1. Vienna E. Brunt et al., "Suppression of the Gut Microbiome Ameliorates Age-Related Arterial Dysfunction and Oxidative Stress in Mice," *Journal of Physiology* 597, no. 9 (May 2019): 2361–78, https://doi.org/10.1113/JP277336.

2. Ron Sender, Shai Fuchs, and Ron Milo, "Revised Estimates for the Number of Human and Bacteria Cells in the Body," *PLoS Biology* 14, no. 8 (August 19, 2016): e1002533, https://doi.org/10.1371/journal.pbio.1002533.

3. Jeremy E. Koenig et al., "Succession of Microbial Consortia in the Developing Infant Gut Microbiome," *Proceedings of the National Academy of Sciences of the USA* 108, Supplement 1 (March 15, 2011): 4578–85, https://doi.org/10.1073/pnas.1000081107.

4. Martin J. Wolff, Mara J. Broadhurst, and Png Loke, "Helminthic Therapy: Improving Mucosal Barrier Function," *Trends in Parasitology* 28, no. 5 (May 2012): 187–94, https://doi.org/10.1016/j.pt.2012.02.008.

5. Helena Helmby, "Human Helminth Therapy to Treat Inflammatory Disorders—Where Do We Stand?," *BMC Immunology* 16, no. 12 (March 26, 2015), https://doi.org/10.1186/s12865-015-0074-3.

6. Grace Rattue, "Autoimmune Disease Rates Increasing," Medical News Today, June 22, 2012, https://www.medicalnewstoday.com/articles/246960.php.

7. Mitsuharu Matsumoto et al., "Longevity in Mice Is Promoted by Probiotic-Induced Suppression of Colonic Senescence Dependent on Upregulation of Gut Bacterial Polyamine Production," *PLoS One* 6, no. 8 (2011): e23652, https://doi.org/10.1371/journal.pone.0023652.

8. Maria G. Dominguez-Bello et al., "Delivery Mode Shapes the Acquisition and Structure of the Initial Microbiota Across Multiple Body Habitats in

Newborns," *Proceedings of the National Academy of Sciences of the USA* 107 (June 29, 2010): 11971–75, https://doi.org/10.1073/pnas.1002601107.

9. Prescilla V. Jeurink et al., "Human Milk: A Source of More Life Than We Imagine," *Beneficial Microbes* 4, no. 1 (March 2013): 17–30, https://doi.org /10.3920/BM2012.0040.

10. Meghan B. Azad et al., "Gut Microbiota of Healthy Canadian Infants: Profiles by Mode of Delivery and Infant Diet at 4 Months," *Canadian Medical Association Journal* 185, no. 5 (March 19, 2013): 385–94, https://doi .org/10.1503/cmaj.121189.

11. Koenig et al., "Succession of Microbial Consortia."

12. Quang N. Nguyen et al., "The Impact of the Gut Microbiota on Humoral Immunity to Pathogens and Vaccination in Early Infancy," *PLoS Pathogens* 12, no. 2 (December 2016): e1005997, https://doi.org/10.1371/journal. ppat.1005997.

13. Evalotte Decker, Mathias Hornef, and Silvia Stockinger, "Cesarean Delivery Is Associated with Celiac Disease but Not Inflammatory Bowel Disease in Children," *Gut Microbes* 2 (2011): 91–98, https://doi.org/10.4161 /gmic.2.2.15414.

14. Amy Langdon, Nathan Crook, and Gautam Dantas, "The Effects of Antibiotics on the Microbiome Throughout Development and Alternative Approaches for Therapeutic Modulation," *Genome Medicine* 8 (2016): 39, https://doi.org/10.1186/s13073-016-0294-z.

15. Robert J. Ferrante et al., "Histone Deacetylase Inhibition by Sodium Butyrate Chemotherapy Ameliorates the Neurodegenerative Phenotype in Huntington's Disease Mice," *Journal of Neuroscience* 23, no. 28 (October 15, 2003): 9418–27, https://doi.org/10.1523/JNEUROSCI.23-28-09418 .2003.

16. Mingyao Ying et al., "Sodium Butyrate Ameliorates Histone Hypoacetylation and Neurodegenerative Phenotypes in a Mouse Model for DRPLA," *Journal of Biological Chemistry* 281, no. 18 (May 5, 2006): 12580–86, https://doi.org/10.1074/jbc.M511677200.

17. Will Chu, "Review Reiterates Fibre's Prebiotic Benefits in Warding Off Stroke and Diabetes," NUTRAingredients.com, January 11, 2019, https://www .nutraingredients.com/Article/2019/01/09/Review-reiterates-fibre-s -prebiotic-benefits-in-warding-off-stroke-and-diabetes.

18. Katie A. Meyer et al., "Carbohydrates, Dietary Fiber, and Incident Type 2 Diabetes in Older Women," *American Journal of Clinical Nutrition* 71, no. 4 (April 2000): 921–30, https://doi.org/10.1093/ajcn/71.4.921.

19. Yikyung Park et al., "Dietary Fiber Intake and Risk of Breast Cancer in Postmenopausal Women: The National Institutes of Health–AARP Diet and Health Study," *American Journal of Clinical Nutrition* 90, no. 3 (September 2009): 664–71, https://doi.org/10.3945/ajcn.2009.27758.

20. James M. Lattimer and Mark D. Haub, "Effects of Dietary Fiber and Its Components on Metabolic Health," *Nutrients* 2, no. 12 (December 2010): 1266–89, https://doi.org/10.3390/nu2121266.

21. Chunye Chen et al., "Therapeutic Effects of Soluble Dietary Fiber Consumption on Type 2 Diabetes Mellitus," *Experimental and Therapeutic Medicine* 12, no. 2 (August 2016): 1232–42, https://doi.org/10.3892/etm.2016.3377.

22. Chen et al., "Therapeutic Effects."

23. Karin de Punder and Leo Pruimboom, "The Dietary Intake of Wheat and Other Cereal Grains and Their Role in Inflammation," *Nutrients* 5, no. 3 (2013): 771–87, https://doi.org/10.3390/nu5030771.

24. A. Pusztai et al., "Antinutritive Effects of Wheat-Germ Agglutinin and Other N-Acetylglucosamine-Specific Lectins," *British Journal of Nutrition* 70, no. 1 (July 1993): 313–21, https://doi.org/10.1079/BJN19930124.

25. Martinette T. Streppel et al., "Dietary Fiber Intake in Relation to Coronary Heart Disease and All-Cause Mortality over 40 y: The Zutphen Study," *American Journal of Clinical Nutrition* 88, no. 4 (October 2008): 1119–25, https://doi.org/10.1093/ajcn/88.4.1119.

26. Park et al., "Dietary Fiber Intake."

27. Diane E. Threapleton et al., "Dietary Fibre Intake and Risk of Cardiovascular Disease: Systematic Review and Meta-Analysis," *BMJ* 347 (December 19, 2013): f6879, https://doi.org/10.1136/bmj.f6879.

28. David L. Topping, Michihiro Fukushima, and Anthony R. Bird, "Resistant Starch as a Prebiotic and Synbiotic: State of the Art," *Proceedings of the Nutrition Society* 62, no. 1 (February 2003): 171–76, https://doi.org/10.1079/PNS2002224.

29. Akbar Aliasgharzadeh et al., "Resistant Dextrin, as a Prebiotic, Improves Insulin Resistance and Inflammation in Women with Type 2 Diabetes: A Randomised Controlled Clinical Trial," *British Journal of Nutrition* 113, no. 2 (January 28, 2015): 321–30, https://doi.org/10.1017/S0007114514003675.

30. University of Colorado Denver, "Diet of Resistant Starch Helps the Body Resist Colorectal Cancer," ScienceDaily, February 19, 2013, www.sciencedaily.com/releases/2013/02/130219140716.htm.

31. Kevin C. Maki et al., "Resistant Starch from High-Amylose Maize Increases Insulin Sensitivity in Overweight and Obese Men," *Journal of Nutrition* 142, no. 4 (April 2012): 717–23, https://doi.org/10.3945/jn.111.152975.

32. Christopher L. Gentile et al., "Resistant Starch and Protein Intake Enhances Fat Oxidation and Feelings of Fullness in Lean and Overweight/Obese Women," *Nutrition Journal* 14 (October 29, 2015): 113, https://doi.org/10.1186/s12937-015-0104-2.

33. Akira Andoh et al., "Comparison of the Gut Microbial Community Between Obese and Lean Peoples Using 16S Gene Sequencing in a Japanese

Population," *Journal of Clinical Biochemistry and Nutrition* 59, no. 1 (July 2016): 65–70, https://doi.org/10.3164/jcbn.15-152.

34. Andoh et al., "Comparison."

35. Peter J. Turnbaugh et al., "A Core Gut Microbiome in Obese and Lean Twins," *Nature* 457, no. 7228 (January 22, 2009): 480–84, https://doi.org/10.1038/nature07540.

36. Saskia Van Hemert et al., "The Role of the Gut Microbiota in Mood and Behavior. Whether Psychobiotics Can Become an Alternative in Therapy in Psychiatry?," *European Psychiatry* 33, Supplement (March 2016): S26, https://doi.org/10.1016/j.eurpsy.2016.01.842.

37. Alessio Fasano, "Leaky Gut and Autoimmune Diseases," *Clinical Reviews in Allergy and Immunology* 42, no. 1 (February 2012): 71–78, https://doi.org/10.1007/s12016-011-8291-x.

38. Bjoern O. Schroeder et al., "Bifidobacteria or Fiber Protects Against Diet-Induced Microbiota-Mediated Colonic Mucus Deterioration," *Cell Host & Microbe* 23, no. 1 (January 10, 2018): P27–40, https://doi.org/10.1016/j.chom.2017.11.004.

39. Van Hemert et al., "Role of the Gut Microbiota."

40. Alper Evrensel and Mehmet Emin Ceylan, "The Gut-Brain Axis: The Missing Link in Depression," *Clinical Psychopharmacology and Neuroscience* 13, no. 3 (December 31, 2015): 239–44, https://doi.org/10.9758/cpn.2015.13.3.239.

41. Andrew H. Moeller et al., "Social Behavior Shapes the Chimpanzee Pan-Microbiome," *Science Advances* 2, no. 1 (January 15, 2016): e1500997, https://doi.org/10.1126/sciadv.1500997.

42. James Gallagher, "How Bacteria Are Changing Your Mood," BBC News, April 24, 2018, https://www.bbc.com/news/health-43815370.

43. Kirsten Tillisch et al., "Brain Structure and Response to Emotional Stimuli as Related to Gut Microbial Profiles in Healthy Women," *Psychosomatic Medicine* 79, no. 8 (October 2017): 905–13, https://doi.org/10.1097/PSY.0000000000000493.

44. Michael T. Bailey et al., "Exposure to a Social Stressor Alters the Structure of the Intestinal Microbiota: Implications for Stressor-Induced Immunomodulation," *Brain, Behavior, and Immunity* 25, no. 3 (March 2011): 397–407, https://doi.org/10.1016/j.bbi.2010.10.023.

45. Peter C. Konturek, Thomas Brzozowski, and S. J. Konturek, "Stress and the Gut: Pathophysiology, Clinical Consequences, Diagnostic Approach and Treatment Options," *Journal of Physiology and Pharmacology* 62, no. 6 (December 2011): 591–99, https://www.ncbi.nlm.nih.gov/pubmed/22314561.

46. Martin F. Graham et al., "Collagen Synthesis by Human Intestinal Smooth Muscle Cells in Culture," *Gastroenterology* 92, no. 2 (February 1987): 400–05, https://doi.org/10.1016/0016-5085(87)90134-X.

PART III: HEAL LIKE A DEITY
1. Vanessa McMains, "Johns Hopkins Study Suggests Medical Errors Are the Third Leading Cause of Death in U.S," Hub, Johns Hopkins University, May 3, 2016, https://hub.jhu.edu/2016/05/03/medical-errors-third-leading-cause-of-death/.

CHAPTER 12: VIRGIN CELLS AND VAMPIRE BLOOD
1. Karen L. Herbst and Thomas Rutledge, "Pilot Study: Rapidly Cycling Hypobaric Pressure Improves Pain After 5 Days in Adiposis Dolorosa," *Journal of Pain Research* 3 (August 20, 2010): 147–53, https://doi.org/10.2147/JPR.S12351.

2. Rex E. Newnham, "Essentiality of Boron for Health Bones and Joints," *Environmental Health Perspectives* 102, Supplement 7 (November 1994): 83–85, https://doi.org/10.1289/ehp.94102s783.

3. Selami Demirci et al., "Boron Increases the Cell Viability of Mesenchymal Stem Cells After Long-Term Cryopreservation," *Cryobiology* 68, no. 1 (February 2014): 139–46, https://doi.org/10.1016/j.cryobiol.2014.01.010.

4. George Dan Mogoșanu et al., "Calcium Fructoborate for Bone and Cardiovascular Health," *Biological Trace Element Research* 172, no. 2 (August 2016): 277–81, https://doi.org/10.1007/s12011-015-0590-2; Zbigniew Pietrzkowski et al., "Short-Term Efficacy of Calcium Fructoborate on Subjects with Knee Discomfort: A Comparative, Double-Blind, Placebo-Controlled Clinical Study," *Clinical Interventions in Aging* 9 (June 5, 2014): 895–99, https://doi.org/10.2147/CIA.S64590.

5. Ezgi Avşar Abdik et al., "Suppressive Role of Boron on Adipogenic Differentiation and Fat Deposition in Human Mesenchymal Stem Cells," *Biological Trace Element Research* 188, no. 2 (April 2019): 384–92, https://doi.org/10.1007/s12011-018-1428-5.

6. Anne Trafton, "Fasting Boosts Stem Cells' Regenerative Capacity," MIT News, May 3, 2018, http://news.mit.edu/2018/fasting-boosts-stem-cells-regenerative-capacity-0503.

7. Massimiliano Cerletti et al., "Short-Term Calorie Restriction Enhances Skeletal Muscle Stem Cell Function," *Cell Stem Cell* 10, no. 5 (May 4, 2012): P515–519, https://doi.org/10.1016/j.stem.2012.04.002.

8. Ting Lo et al., "Glucose Reduction Prevents Replicative Senescence and Increases Mitochondrial Respiration in Human Mesenchymal Stem Cells," *Cell Transplantation* 30, no. 6 (2011): 813–25, https://doi.org/10.3727/096368910X539100.

9. Maria Carmen Valero et al., "Eccentric Exercise Facilitates Mesenchymal Stem Cell Appearance in Skeletal Muscle," *PLoS One* 7, no. 1 (January 11, 2012): e29760, https://doi.org/10.1371/journal.pone.0029760.

10. Joerg Hucklenbroich et al., "Aromatic-Turmerone Induces Neural Stem Cell Proliferation *in vitro* and *in vivo*," *Stem Cell Research & Therapy* 5, no. 4 (September 26, 2014): 100, https://doi.org/10.1186/scrt500.

11. Dong Suk Yoon et al., "SIRT1 Directly Regulates SOX2 to Maintain Self-Renewal and Multipotency in Bone Marrow-Derived Mesenchymal Stem Cells," *Stem Cells* 32, no. 12 (December 2014): 3219–31, https://doi.org/10.1002/stem.1811.

12. "Natural Ways to Increase Stem Cell Activity," Stem Cell The Magazine, October 18, 2017, https://stemcellthemagazine.com/2017/10/natural-ways-to-increase-stem-cell-activity/.

13. Tsung-Jung Ho et al., "Tai Chi Intervention Increases Progenitor CD34(+) Cells in Young Adults," *Cell Transplantation* 23, no. 4–5 (2014): 613–20, https://doi.org/10.3727/096368914X678355.

14. Koh, "A Good Night's Sleep Keeps Your Stem Cells Young," dkfz (Deutsches Krebsforschungszentrum), February 18, 2015, https://www.dkfz.de/en/presse/pressemitteilungen/2015/dkfz-pm-15-08-A-good-nights-sleep-keeps-your-stem-cells-young.php; Hoda Elkhenany, "Tissue Regeneration: Impact of Sleep on Stem Cell Regenerative Capacity," *Life Sciences* 214 (December 1, 2018): 51–61, https://doi.org/10.1016/j.lfs.2018.10.057.

15. Ilan Gruenwald et al., "Shockwave Treatment of Erectile Dysfunction," *Therapeutic Advances in Urology* 5, no. 2 (April 2013): 95–9, https://doi.org/10.1177/1756287212470696.

16. Michaela Z. Ratajczak et al., "Very Small Embryonic-Like Stem Cells (VSELs) Represent a Real Challenge in Stem Cell Biology: Recent Pros and Cons in the Midst of a Lively Debate," *Leukemia* 28 (2014): 473–84, https://doi.org/10.1038/leu.2013.255.

17. Diane M. Jaworski and Leonor Pérez-Martínez, "Tissue Inhibitor of Metalloproteinase-2 (TIMP-2) Expression Is Regulated by Multiple Neural Differentiation Signals," *Journal of Neurochemistry* 98, no. 1 (July 2006): 234–47, https://doi.org/10.1111/j.1471-4159.2006.03855.x.

18. Ye Li et al., "Human iPSC-Derived Natural Killer Cells Engineered with Chimeric Antigen Receptors Enhance Anti-Tumor Activity," *Cell Stem Cell* 23, no. 2 (August 2, 2018): P181–192.E5, https://doi.org/10.1016/j.stem.2018.06.002.

19. Rich Haridy, "Anti-Aging Discovery Reveals Importance of Immune System in Clearing Old Cells," New Atlas, January 1, 2019, https://newatlas.com/immune-system-aging-senescent-cells/57835/.

20. "Natural Killer Cell," ScienceDaily, https://www.sciencedaily.com/terms/natural_killer_cell.htm.

21. Qing Li et al., "Effect of Phytoncide from Trees on Human Natural Killer Cell Function," *International Journal of Immunopathology and Pharmacology* 22, no. 4 (October–December 2009): 951–59, https://doi.org/10.1177/039463200902200410.

22. Ebere Anyanwu et al., "The Neurological Significance of Abnormal Natural Killer Cell Activity in Chronic Toxigenic Mold Exposures," *Scientific World Journal* 13, no. 3 (November 13, 2003): 1128–37, https://doi.org/10.1100/tsw.2003.98.

23. Saul E. Villeda et al., "Young Blood Reverses Age-Related Impairments in Cognitive Function and Synaptic Plasticity in Mice," *Nature Medicine* 20 (2014): 659–63, https://doi.org/10.1038/nm.3569.

24. Makoto Kuro-o et al., "Mutation of the Mouse Klotho Gene Leads to a Syndrome Resembling Ageing," *Nature* 390, no. 6655 (November 6, 1997): 45–51, https://doi.org/10.1038/36285.

25. Hiroshi Kurosu et al., "Suppression of Aging in Mice by the Hormone Klotho," *Science* 309, no. 5742 (September 16, 2005): 1829–33, https://doi.org/10.1126/science.1112766.

26. Richard D. Semba et al., "Plasma Klotho and Mortality Risk in Older Community-Dwelling Adults," *Journals of Gerontology Series A: Biological Sciences & Medical Sciences* 66, no. 7 (July 2011): 794–800, https://doi.org/10.1093/gerona/glr058.

27. Dan E. Arking et al., "Association of Human Aging with a Functional Variant of Klotho," *Proceedings of the National Academy of Sciences of the USA* 99, no. 2 (January 2002): 856–61, https://doi.org/10.1073/pnas.022484299.

28. Jennifer S. Yokoyama et al., "Variation in Longevity Gene KLOTHO Is Associated with Greater Cortical Volumes," *Annals of Clinical and Translational Neurology* 2, no. 3 (January 2015): 215–30, https://doi.org/10.1002/acn3.161.

29. Ming-Chang Hu et al., "Klotho Deficiency Is an Early Biomarker of Renal Ischemia-Reperfusion Injury and Its Replacement Is Protective," *Kidney International* 78, no. 12 (December 2010): 1240–51, https://doi.org/10.1038/ki.2010.328; Ming-Chang Hu et al., "Recombinant α-Klotho May Be Prophylactic and Therapeutic for Acute to Chronic Kidney Disease Progression and Uremic Cardiomyopathy," *Kidney International* 91, no. 5 (January 2017): 1104–14, https://doi.org/10.1016/j.kint.2016.10.034.

30. Richard D. Semba et al., "Klotho in the Cerebrospinal Fluid of Adults With and Without Alzheimer's Disease," *Neuroscience Letters* 558 (January 2014): 37–40, https://doi.org/10.1016/j.neulet.2013.10.058.

31. Julio Leon et al., "Peripheral Elevation of a Klotho Fragment Enhances Brain Function and Resilience in Young, Aging and Alpha-Synuclein Transgenic Mice," *Cell Reports* 20: 1360–71, https://doi.org/10.1016/j.celrep.2017.07.024.

32. Shigehiro Doi et al., "Klotho Inhibits Transforming Growth Factor-β1 (TGF-β1) Signaling and Suppresses Renal Fibrosis and Cancer Metastasis in Mice," *Journal of Biological Chemistry* 286, no. 10 (March 11, 2011): 8655–65, https://doi.org/10.1074/jbc.M110.174037.

33. Elisabete A. Forsberg et al., "Effect of Systemically Increasing Human Full-Length Klotho on Glucose Metabolism in db/db Mice," *Diabetes Research and Clinical Practice* 113 (March 2016): 208–10, https://doi.org/10.1016/j.diabres.2016.01.006.

34. Richard D. Semba et al., "Relationship of Low Plasma Klotho with Poor Grip Strength in Older Community-Dwelling Adults: The InCHIANTI

Study," *European Journal of Applied Physiology* 112, no. 4 (April 2012): 1215–20, https://www.ncbi.nlm.nih.gov/pubmed/21769735.

35. Lisa D. Chong, "Repairing Injured Muscle," *Science*, December 14, 2018, http://science.sciencemag.org/content/362/6420/1260.5.full.

36. Morgan S. Saghiv et al., "The Effects of Aerobic and Anaerobic Exercise on Circulating Soluble-Klotho and IGF-1 in Young and Elderly Adults and in CAD Patients," *Journal of Circulating Biomarkers* 6 (September 28, 2017): 6:1849454417733388, https://doi.org/10.1177/1849454417733388.

37. Wei Ling Lau et al., "Vitamin D Receptor Agonists Increase Klotho and Osteopontin While Decreasing Aortic Calcification in Mice with Chronic Kidney Disease Fed a High Phosphate Diet," *Kidney International* 82, no. 12 (December 2012): 1261–70, https://doi.org/10.1038/ki.2012.322.

38. Hye Eun Yoon et al., "Angiotensin II Blockade Upregulates the Expression of Klotho, the Anti-Ageing Gene, in an Experimental Model of Chronic Cyclosporine Nephropathy," *Nephrology Dialysis Transplantation* 26, no. 3 (March 2011): 800–13, https://doi.org/10.1093/ndt/gfq537.

39. Shih-Che Hsu et al., "Testosterone Increases Renal Anti-Aging Klotho Gene Expression via the Androgen Receptor-Mediated Pathway," *Biochemical Journal* 464, no. 2 (December 2014): 221–29, https://doi.org/10.1042/BJ20140739.

40. Gerit D. Mulder et al., "Enhanced Healing of Ulcers in Patients with Diabetes by Topical Treatment with Glycyl-L-Histidyl-L-Lysine Copper," *Wound Repair and Regeneration* 2, no. 4 (October 1994): 259–69, https://doi.org/10.1046/j.1524-475X.1994.20406.x.

41. Loren Pickart, Jessica Michelle Vasquez-Soltero, and Anna Margolina, "The Human Tripeptide GHK-Cu in Prevention of Oxidative Stress and Degenerative Conditions of Aging: Implications for Cognitive Health," *Oxidative Medicine and Cellular Longevity* 2012 (February 2012): 324832, https://doi.org/10.1155/2012/324832.

42. Loren Pickart, "The Human Tri-Peptide GHK and Tissue Remodeling," *Journal of Biomaterials Science, Polymer Edition* 19, no. 8 (2008): 969–88, https://doi.org/10.1163/156856208784909435.

43. Mary P. Lupo and Anna L. Cole, "Cosmeceutical Peptides," *Dermatologic Therapy* 20, no. 5 (November 28, 2007): 343–49, https://doi.org/10.1111/j.1529-8019.2007.00148.x.

44. Loren Pickart, Jessica Michelle Vasquz-Soltero, and Anna Margolina, "GHK Peptide as a Natural Modulator of Multiple Cellular Pathways in Skin Regeneration," *BioMed Research International* 2015 (April 2015): 648108, http://dx.doi.org/10.1155/2015/648108.

CHAPTER 13: DON'T LOOK LIKE AN ALIEN: AVOIDING BALDNESS, GRAYS, AND WRINKLES

1. Nicole Verzijl et al., "Effect of Collagen Turnover on the Accumulation of Advanced Glycation End Products," *Journal of Biological Chemistry* 275 (December 15, 2000): 39027–31, http://doi.org/10.1074/jbc.M006700200.

2. Ruta Ganceviciene et al., "Skin Anti-Aging Strategies," *Dermatoendocrinology* 4, no. 3 (2012): 308–19, http://doi.org/10.4161/derm.22804.

3. Ketavan Jariashvili et al., "UV Damage of Collagen: Insights from Model Collagen Peptides," *Biopolymers* 97, no. 3 (March 2012): 189–98, http://doi.org/10.1002/bip.21725; A. Knuutinen et al., "Smoking Affects Collagen Synthesis and Extracellular Matrix Turnover in Human Skin," *British Journal of Dermatology* 146, no. 4 (April 2002): 588–94, https://doi.org/10.1046/j.1365-2133.2002.04694.x.

4. Ehrhardt Proksch et al., "Oral Intake of Specific Bioactive Collagen Peptides Reduces Skin Wrinkles and Increases Dermal Matrix Synthesis," *Skin Pharmacology and Physiology* 27, no. 3 (2014): 113–19, https://doi.org/10.1159/000355523; Ehrhardt Proksch et al., "Oral Supplementation of Specific Collagen Peptides Has Beneficial Effects on Human Skin Physiology: A Double-Blind, Placebo-Controlled Study," *Skin Pharmacology and Physiology* 27, no. 1 (2014): 47–55, https://doi.org/10.1159/000351376.

5. Kristine L. Clark et al., "24-Week Study on the Use of Collagen Hydrolysate as a Dietary Supplement in Athletes with Activity-Related Joint Pain," *Current Medical Research and Opinion* 24, no. 5 (May 2008): 1485–96, https://doi.org/10.1185/030079908X291967.

6. Olivier Bruyère et al., "Effect of Collagen Hydrolysate in Articular Pain: A 6-Month Randomized, Double-Blind, Placebo Controlled Study," *Complementary Therapies in Medicine* 20, no. 3 (June 2012): 124–30, https://doi.org/10.1016/j.ctim.2011.12.007.

7. Daniel König et al., "Specific Collagen Peptides Improve Bone Mineral Density and Bone Markers in Postmenopausal Women—A Randomized Controlled Study," *Nutrients* 10, no. 1 (January 2018): E97, https://doi.org/10.3390/nu10010097.

8. Martin F. Graham et al., "Collagen Synthesis by Human Intestinal Smooth Muscle Cells in Culture," *Gastroenterology* 92, no. 2 (February 1987): 400–05, https://www.ncbi.nlm.nih.gov/pubmed/3792777.

9. Kenji Nagahama et al., "Orally Administered L-Arginine and Glycine Are Highly Effective Against Acid Reflux Esophagitis in Rats," *Medical Science Monitor* 18, no. 1 (2012): BR9–15, https://doi.org/10.12659/MSM.882190.

10. James English, "Gastric Balance: Heartburn Not Always Caused by Excess Acid," *Nutrition Review*, November 25, 2018, https://nutritionreview.org/2018/11/gastric-balance-heartburn-caused-excess-acid/.

11. Morton I. Grossman, Joseph B. Kirsner, and Ian E. Gillespie, "Basal and Histalog-Stimulated Gastric Secretion in Control Subjects and in Patients with Peptic Ulcer or Gastric Cancer," *Gastroenterology* 45 (July 1963): 15–26, https://doi.org/10.1016/S0016-5085(19)34918-2.

12. Stephen D. Krasinski et al., "Fundic Atrophic Gastritis in an Elderly Population. Effect on Hemoglobin and Several Serum Nutritional Indicators,"

Journal of the American Geriatric Society 34, no. 11 (November 1986): 800–06, https://doi.org/10.1111/j.1532-5415.1986.tb03985.x.

13. Wataru Yamadera et al., "Glycine Ingestion Improves Subjective Sleep Quality in Human Volunteers, Correlating with Polysomnographic Changes," *Sleep and Biological Rhythms* 5, no. 2 (April 2007): 126–31, https://doi.org/10.1111/j.1479-8425.2007.00262.x.

14. Edward D. Harris Jr. and Peter A. McCroskery, "The Influence of Temperature and Fibril Stability on Degradation of Cartilage Collagen by Rheumatoid Synovial Collagenase," *New England Journal of Medicine* 290 (January 1974): 1–6, https://doi.org/10.1056/NEJM197401032900101; Harris and McCroskery, "Influence."

15. Anna Lubkowska, Barbara Dołęgowska, and Zbigniew Szyguła, "Whole-Body Cryostimulation—Potential Beneficial Treatment for Improving Antioxidant Capacity in Healthy Men—Significance of the Number of Sessions," *PLoS One* 7, no. 10 (October 15, 2012): e46352, https://doi.org/10.1371/journal.pone.0046352.

16. Imran Majid, "Microneedling Therapy in Atrophic Facial Scars: An Objective Assessment," *Journal of Cutaneous and Aesthetic Surgery* 2, no. 1 (2009): 26–30, https://doi.org/10.4103/0974-2077.53096.

17. Simran Chawla, "Split Face Comparative Study of Microneedling with PRP Versus Microneedling with Vitamin C in Treating Atrophic Post Acne Scars," *Journal of Cutaneous and Aesthetic Surgery* 7, no. 4 (2014): 209–12, https://doi.org/10.4103/0974-2077.150742.

18. Seung-Hye Hong et al., "Alternative Biotransformation of Retinal to Retinoic Acid or Retinol by an Aldehyde Dehydrogenase from *Bacillus cereus*," *Applied and Environmental Microbiology* 82, no. 13 (June 13, 2016), https://doi.org/10.1128/AEM.00848-16.

19. Rong Kong et al., "A Comparative Study of the Effects of Retinol and Retinoic Acid on Histological, Molecular, and Clinical Properties of Human Skin," *Journal of Cosmetic Dermatology* 15, no. 1 (March 2016): 49–57, https://doi.org/10.1111/jocd.12193.

20. Pierpaolo Mastroiacovo et al., "High Vitamin A Intake in Early Pregnancy and Major Malformations: A Multicenter Prospective Controlled Study," *Teratology* 59, no. 1 (January 1999): 7–11, https://doi.org/10.1002/(SICI)1096-9926(199901)59:1<7::AID-TERA4>3.0.CO;2-6.

21. Ratan K. Chaudhuri and Krzysztof Bojanowski, "Bakuchiol: A Retinol-Like Functional Compound Revealed by Gene Expression Profiling and Clinically Proven to Have Anti-Aging Effects," *International Journal of Cosmetic Science* 36, no. 3 (June 2014): 221–30, https://doi.org/10.1111/ics.12117.

22. Zheng-Mei Xiong et al., "Anti-Aging Potentials of Methylene Blue for Human Skin Longevity," *Scientific Reports* 7 (2017): 2475, https://doi.org/10.1038/s41598-017-02419-3.

23. John W. Haycock et al., "α-Melanocyte-Stimulating Hormone Inhibits NF-κB Activation in Human Melanocytes and Melanoma Cells," *Journal of Investigative Dermatology* 113, no. 4 (October 1999): 560–66, https://doi .org/10.1046/j.1523-1747.1999.00739.x.

24. Arturo Solis Herrera and Paola E. Solis Arias, "Einstein Cosmological Constant, the Cell, and the Intrinsic Property of Melanin to Split and Re-Form the Water Molecule," *MOJ Cell Science & Report* 1, no. 2 (August 27, 2014): 46–51, https://doi.org/10.15406/mojcsr.2014.01.00011.

25. Federation of American Societies for Experimental Biology, "Why Hair Turns Gray Is No Longer a Gray Area: Our Hair Bleaches Itself as We Grow Older," ScienceDaily, February 24, 2009, www.sciencedaily.com /releases/2009/02/090223131123.htm.

26. Edith Lubos, Joseph Loscalzo, and Diane E. Handy, "Glutathione Perox-idase-1 in Health and Disease: From Molecular Mechanisms to Thera-peutic Opportunities," *Antioxidants & Redox Signaling* 15, no. 7 (October 2011): 1957–97, https://doi.org/10.1089/ars.2010.3586.

27. Ajay Pal et al., "Ashwagandha Root Extract Inhibits Acetylcholine Ester-ase, Protein Modification and Ameliorates H_2O_2-Induced Oxidative Stress in Rat Lymphocytes," *Pharmacognosy Journal* 9, no. 3 (May–June 2017): 302–09, https://doi.org/10.5530/pj.2017.3.52/.

28. Lakshmi-Chandra Mishra, Betsy B. Singh, and Simon Dagenais, "Scien-tific Basis for the Therapeutic Use of *Withania somnifera* (Ashwagandha): A Review," *Alternative Medicine Review* 5, no. 4 (2000): 334–46, http ://altmedrev.com/archive/publications/5/4/334.pdf.

29. Melissa L. Harris et al., "A Direct Link Between MITF, Innate Immu-nity, and Hair Graying," *PLoS Biology* 16, no. 5 (May 3, 2018): e2003648, https://doi.org/10.1371/journal.pbio.2003648.

30. Thomas Rhodes et al., "Prevalence of Male Pattern Hair Loss in 18–49 Year Old Men," *Dermatologic Surgery* 24, no. 12 (December 1998): 13330–32, https://doi.org/10.1111/j.1524-4725.1998.tb00009.x.

31. Paulo Müller Ramos and Hélio Amante Miot, "Female Pattern Hair Loss: A Clinical and Pathophysiological Review," *Brazilian Annals of Dermatology (Anais Brasileiros de Dermatologia)* 90, no. 4 (July–August 2015): 529–43, https://doi.org/10.1590/abd1806-4841.20153370.

32. Peter Dockrill, "'Unprecedented' DNA Discovery Reverses Wrinkles and Hair Loss in Mice," Science Alert, July 28, 2018, https://www.science alert.com/unprecedented-dna-discovery-actually-reverses-wrinkles-and -hair-loss-mitochondria-mutation-mtdna/amp.

33. Michael P. Zimber et al., "Hair Regrowth Following a Wnt- and Follistatin Containing Treatment: Safety and Efficacy in a First-in-Man Phase 1 Clin-ical Trial," *Journal of Drugs in Dermatology* 20, no. 11 (November 2011): 1308–12, https://www.ncbi.nlm.nih.gov/m/pubmed/22052313/.

34. Zhuo-ming Li, Suo-wen Xu, and Pei-qing Liu, "*Salvia miltiorrhiza*Burge (Danshen): A Golden Herbal Medicine in Cardiovascular Therapeutics,"

Acta Pharmacologica Sinica 39, no. 5 (May 2018): 802–24, https://doi .org/10.1038/aps.2017.193.

35. Martin I. Surks and Laura Boucai, "Age- and Race-Based Serum Thyrotropin Reference Limits," *Journal of Clinical Endocrinology & Metabolism* 95, no. 2 (February 1, 2010): 496–502, https://doi.org/10.1210/jc.2009–1845.

36. Surks and Boucai, "Age- and Race-Based Serum."

37. Susan Jobling et al., "A Variety of Environmentally Persistent Chemicals, Including Some Phthalate Plasticizers, Are Weakly Estrogenic," *Environmental Health Perspectives* 103, no. 6 (June 1995): 582–87, https://doi .org/10.1289/ehp.95103582.

CHAPTER 14: HACK YOUR LONGEVITY LIKE A RUSSIAN

1. Gabriel Sosne, Ping Qiu, and Michelle Kurpakus-Wheater, "Thymosin Beta 4: A Novel Corneal Wound Healing and Anti-Inflammatory Agent," *Clinical Ophthalmology* 1, no. 3 (2007): 201–07, https://www.ncbi.nlm.nih .gov/pmc/articles/PMC2701135/.

2. Chuanyu Wei et al., "Thymosin Beta 4 Protects Mice from Monocrotaline-Induced Pulmonary Hypertension and Right Ventricular Hypertrophy," *PLoS One* 9, no. 11 (November 20, 2014): e110598, https://doi .org/10.1371/journal.pone.0110598.

3. Vladimir Kh. Khavinson and Vyacheslav G. Morozov, "Peptides of Pineal Gland and Thymus Prolong Human Life," *Neuroendocrinology Letters* 24, no. 3 (June–August 2003): 233–40, https://www.ncbi.nlm.nih.gov /pubmed/14523363.

4. Chung-Hsun Chang et al., "The Promoting Effect of Pentadecapeptide BPC 157 on Tendon Healing Involves Tendon Outgrowth, Cell Survival, and Cell Migration," *Journal of Applied Physiology* 110, no. 3 (March 2011): 774–80, https://doi.org/10.1152/japplphysiol.00945.2010.

5. Božidar Šebečić et al., "Osteogenic Effect of a Gastric Pentadecapeptide, BPC-157, on the Healing of Segmental Bone Defect in Rabbits: A Comparison with Bone Marrow and Autologous Cortical Bone Implantation," *Bone* 24, no. 3 (1999): 195–202, https://doi.org/10.1016/S8756 -3282(98)00180-X.

6. Predrag Sikirić et al., "Toxicity by NSAIDs. Counteraction by Stable Gastric Pentadecapeptide BPC 157," *Current Pharmaceutical Design* 19, no. 1 (2013): 76–83, https://www.ncbi.nlm.nih.gov/pubmed/22950504.

7. Tihomir Vuksic et al., "Stable Gastric Pentadecapeptide BPC 157 in Trials for Inflammatory Bowel Disease (PL-10, PLD-116, PL 14736, Pliva, Croatia) Heals Ileoileal Anastomosis in the Rat," *Surgery Today* 37, no. 9 (2007): 768–77, https://doi.org/10.1007/s10787-006-1531-7.

8. Ramesh Narayanan et al., "Selective Androgen Receptor Modulators in Preclinical and Clinical Development," *Nuclear Receptor Signaling* 6 (2008): e010, https://doi.org/10.1621/nrs.06010.

9. Vihang A. Narkar et al., "AMPK and PPARdelta Agonists Are Exercise Mimetics," *Cell* 134, no. 3 (August 2008): 405–15, https://doi.org/10.1016/j.cell.2008.06.051.

10. Weiwei Fan et al., "Road to Exercise Mimetics: Targeting Nuclear Receptors in Skeletal Muscle," *Journal of Molecular Endocrinology* 51, no. 3 (2013): T87–T100, https://doi.org/10.1530/JME-13-0258.

11. Jane A. Mitchell and David Bishop-Bailey, "PPARβ/δ a Potential Target in Pulmonary Hypertension Blighted by Cancer Risk," *Pulmonary Circulation* 9, no. 1 (June–March 2019): 2045894018812053, https://doi.org/10.1177/2045894018812053.

12. Estelle Woldt et al., "Rev-erb-α Modulates Skeletal Muscle Oxidative Capacity by Regulating Mitochondrial Biogenesis and Autophagy," *Nature Medicine* 19, no. 8 (August 2013): 1039–48, https://doi.org/10.1038/nm.3213.

13. Jill P. Smith et al., "Low-Dose Naltrexone Therapy Improves Active Crohn's Disease," *American Journal of Gastroenterology* 102, no. 4 (April 2007): 820–28, https://doi.org/10.1111/j.1572-0241.2007.01045.x.

14. Jarred Younger, Luke Parkitny, and David McLain, "The Use of Low-Dose Naltrexone (LDN) as a Novel Anti-Inflammatory Treatment for Chronic Pain," *Clinical Rheumatology* 33, no. 4 (2014): 451–59, https://doi.org/10.1007/s10067-014-2517-2.

15. Jarred Younger and Sean Mackey, "Fibromyalgia Symptoms Are Reduced by Low-Dose Naltrexone: A Pilot Study," *Pain Medicine* 10, no. 4 (May–June 2009): 663–72, https://doi.org/10.1111/j.1526-4637.2009.00613.x; Jarred Younger et al., "Low-Dose Naltrexone for the Treatment of Fibromyalgia: Findings of a Small, Randomized, Double-Blind, Placebo-Controlled, Counterbalanced, Crossover Trial Assessing Daily Pain Levels," *Arthritis & Rheumatology* 65, no. 2 (February 2013): 529–38, https://doi.org/10.1002/art.37734.

16. Renee N. Donahue, Patricia J. McLaughlin, and Ian S. Zagon, "Low-Dose Naltrexone Suppresses Ovarian Cancer and Exhibits Enhanced Inhibition in Combination with Cisplatin," *Experimental Biology and Medicine (Maywood)* 236, no. 7 (July 2011): 883–95, https://doi.org/10.1258/ebm.2011.011096.

17. Burton M. Berkson, Daniel M. Rubin, and Arthur J. Berkson, "Reversal of Signs and Symptoms of a B-cell Lymphoma in a Patient Using Only Low-Dose Naltrexone," *Integrative Cancer Therapies* 6, no. 3 (September 2007): 293–96, https://doi.org/10.1177/1534735407306358; Ian S. Zagon, Renee N. Donahue, and Patricia J. McLaughlin, "Opioid Growth Factor-Opioid Growth Factor Receptor Axis Is a Physiological Determinant of Cell Proliferation in Diverse Human Cancers," *American Journal of Physiology—Regulatory, Integrative, and Comparative Physiology* 297, no. 4 (October 2009): R1154–61, https://doi.org/10.1152/ajpregu.00414.2009.

18. Gordon L. Cheng et al., "Heroin Abuse Accelerates Biological Aging: A Novel Insight from Telomerase and Brain Imaging Interaction," *Trans-

lational Psychiatry 3, no. 5 (May 21, 2013): e260, https://doi.org/10.1038 /tp.2013.36.

19. Franco Cataldo, "Interaction of C(60) Fullerene with Lipids," *Chemistry and Physics of Lipids* 163, no. 6 (June 2010): 524–29, https://doi.org/10.1016 /j.chemphyslip.2010.03.004.

20. Yuriy Rud et al., "Using C60 Fullerenes for Photodynamic Inactivation of Mosquito Iridescent Viruses," *Journal of Enzyme Inhibition and Medicinal Chemistry* 27, no. 4 (August 2012): 614–17, https://doi.org/10.3109/147563 66.2011.601303.

21. Yuliana Pineda Galvan et al., "Fullerenes as Anti-Aging Antioxidants," *Current Aging Science* 10, no. 1 (2017): 56–67, https://doi.org/10.2174/187460 9809666160921120008.

22. Tarek Baati et al., "The Prolongation of the Lifespan of Rats by Repeated Oral Administration of [60]Fullerene," *Biomaterials* 33, no. 19 (2012): 4936–46, https://doi.org/10.1016/j.biomaterials.2012.03.036.

INDEX

acacia fiber, 195
activated charcoal, 136–38, 143
ActivePQQ, 125
Actos (pioglitazone), 219
ADD (attention deficit disorder), 12–13
Adelson, Harry, 211–15, 217
adenosine triphosphate (ATP), 9, 94, 95, 240
adipose tissue, 52
adsorption, 136
AEP (amino ethanol phosphate), 29, 40
aflatoxin, 137
AGEs (advanced glycation end products), 31–32, 36, 45, 46, 112, 139
agglutinins (WGA), 194–95
aging
  overview, 24
  autoimmune diseases' role in, 188
  autophagy for slowing down, 19
  from birth control pills, 168, 169
  and calcification of tissue, 140–41
  and deep sleep decrease, 70
  and heart rate variability, 72
  and heavy metal toxicity, 134
  hormone level reduction, 158–59
  and leaky gut syndrome, 199
  lessons from Greek myths, xvii, 207, 227, 265
  and telomere length, 37
  See also death by a thousand cuts; Seven Pillars of Aging
aging backward, 204–64
  overview, xviii, 96, 101–2, 232–33, 248

NAD+ to NADH ratio, 152–53
  and peptides, 248–52
  and stem cell therapy, 211
  See also multiple methods for strengthening your body
alchemy, xiv
Alitura skin-care products, 230
alpha-lipoic acid (ALA), 135, 143
alpha-melanocyte-stimulating hormone (alpha-MSH), 240
alpha-MSH (alpha-melanocyte-stimulating hormone), 240
aluminum, 130
Alzheimer's disease
  overview, 17–19
  and blood sugar level, 14, 113–15
  development of, 104–5
  effects of nicotine on, 120
  and heavy metal toxicity, 34
  incidence of, 4
  and insulin, 115
  and Klotho in cerebrospinal fluid, 228
  light therapy for, 111
  prevention, 65, 104–5, 114–15, 128
Amen, Daniel, 17–18
American Academy of Anti-Aging Medicine conference, 169
American Cancer Society, 66–67
amino acids
  overview, 47–48, 248
  EDTA, 139–41
  GHK-Cu, 229–30, 231, 249
  glycine, 43, 234–36

amino acids (*cont.*)
  L-theanine, 125, 128
  L-tyrosine, 171–72
  peptides, 248–52, 263
  vitamin C, collagen, and, 7
amino ethanol phosphate (AEP), 29, 40
amyloid proteins
  overview, 33–34
  curcumin vs., 126
  glymphatic system vs., 65
  and heavy metals, 34–35
  and high sugar diet, 112–13
  light therapy for breaking up, 111
  and microglial cells, 108
  monitoring, 126
  NAD+ for avoiding buildup, 152
  vitamin D vs., 87
anaerobic metabolism, 19
andropause, 159
aniracetam, 118
antibacterial, coconut oil as, 182
antibiotics
  in animals grown for human food, 12,
    47, 48, 201
  and antibiotic-resistant bacteria, 12,
    146, 201–2
  Asprey's use of, in childhood, 3, 185,
    201–2
  and gray hair, 243
  and microbiomes, 185, 187–88, 190
  ozone therapy vs., 144, 146, 147–48
anti-inflammatories
  alpha-MSH hormone, 240
  butyric acid, 51–52
  coconut oil, 182
  omega-3 fats, 53
  short-chain fatty acids, 11–12
antioxidants
  anti-aging properties, 9
  carbon 60, 261
  for catalase production increase, 241
  curcumin, 80, 126–27, 219
  dismutase, 127
  fisetin, 30
  for heavy metal detox, 134–36
  lion's mane mushroom, 125–26
  and metabolites from fasting, 49

methylene blue, 238–39
polyphenols, 30, 40, 59, 197, 201,
    203, 246
PQQ, 124–25
superoxide dismutase, 122–23
vitamin C, 236
See also glutathione
antioxidant therapy via IV, 22
anxiety disorders. *See* stress
apoptosis, 9, 19, 25, 37, 38–39
Apple's Night Shift (warm screen
    colors), 81–82
arsenic, 35, 130–33, 138, 141
arthritis
  prevention, 234
  and senescent cells, 29
  from senescent cells, 29
  from toxic mold exposure, 3
  treatments, 155, 181, 211, 216–17,
    218–19
ashitaba (Japanese herb), 30
ashwagandha, 241
Asperger's syndrome, 12–13
Asprey, Anna, 147–48
Asprey, Dave
  allergy tests reveal mold allergy, 7
  business school, 17–18, 55, 68, 103
  childhood, xix, 3–4
  college years, 4, 15–16, 17
  and Epitalon, 38
  hacking his brain's language
    processing area with a laser (not
    recommended), 110–11
  Hashimoto's thyroiditis, 5–6
  home laboratory
  hormone levels, 5–6, 156
  light therapy for whiplash, 91–92
  living proclamation, 265–66
  meeting his future wife, 169–70
  parasite introduction into his gut, 187
  POTS from toxic mold exposure,
    12–13
  questioning conventional wisdom
    about diet, 14–15
  studying anti-aging results, 6
  Tibet and Nepal trip, 55–57
  on trying new methods, 207–8

Viome microbiome analysis results, 201–2
"zero carbs for nineteen days" experiment, 69
Asprey, Lana
infertility diagnosis and cure, 169–70
stem cell treatments, 213–14, 220
treating Anna's scratched ear, 147–48
utilizing red light panels, 93
asthma, 177
astragalus (herb), 39
atherosclerosis, 10–11, 172
athletes and performance-enhancing substances, 253–54
Atmospheric Cell Trainer, 212–13
ATP. See adenosine triphosphate
attention deficit disorder (ADD), 12–13
autoimmune diseases
and amyloids, 33–34
Hashimoto's thyroiditis, 5–6
and leaky gut syndrome, 198
on modafinil, 118–21
and Naltrexone, 259, 263
prevention, 187–88
and wheat, 42
autonomic nervous system, 71–72
autophagy, 19, 34, 48–49
autophagy stimulation, 22
Ayurvedic medicine, 30, 39, 136, 181–82, 238, 241

bacteria
overview, 185–86
in breast milk, 189
and histamines, 192
human acquisition of, 7
interaction with other bacteria, 11–12
mold as lethal threat to, 7–8
reducing mouth bacteria, 182
See also gut bacteria
bakuchiol, 238
baldness, 242–45
Basis, 68
BDNF (brain-derived neurotrophic factor), 117, 125–26
benzophenones, 244, 245, 246
betaine hydrochloride (HCL), 235

Better Baby Book, The (Asprey and Asprey), 170
bifidobacteria, 198
Big Food companies and glyphosate, 44
biohackers, xiv, 213
biohacking
overview, xvii–xviii, 22
Asprey's home laboratory, 93, 107, 146–48
and hormones, 170–72
with rapamycin, 30
rule 1: remove what makes you weak, 10
See also multiple methods for strengthening your body
bioidentical hormone replacement therapy
Asprey's experience with, 6, 156
DHEA's longevity-enhancing effects, 162–63, 173
human growth hormone, 165–66
medical tests prior to, 160–61, 163
oxytocin, 163–65
pellets put under skin by a doctor, 162
as smart drug, 159
synthetic hormones vs., 160
testosterone, 161–62, 257
for thyroid hormone, 172
traditional view of, 157–58, 159–60
Wiley's research and recommendations, 157, 159
bioregulator peptides, 249–50, 263
birth control pills, 168, 169
bite alignment, 178–80
bite guards, 178–79, 180–81, 184
blackout curtains, 80–81
black pepper extract, 126–27
blood-brain barrier and toxic metals, 130
blood flow and leaky gut syndrome, 198
blood flow and trigeminal nerve, 180
blood pressure and Klotho levels, 228
blood sugar
and AGEs, 31–32, 45, 112
and bad quality sleep, 66
damage caused by, 13–14, 31–32, 112–13
fasting blood sugar test, 5

blood sugar (*cont.*)
  and glucosamine, 15
  monitoring, 113–14
  and Ostarine, 256
  stabilizing your level of, 128
  and sugar cravings, 44–45
  and UVB exposure, 87
blue (junk) light, 78–82, 83, 85–90, 97
bones, keeping calcium in, 140
boron supplements, 219, 231
Boston University, 43, 111–12
BPC157 healing peptide, 252
brain
  blood-brain barrier and toxic
    metals, 130
  blood flow and trigeminal nerve, 180
  gut-brain axis, 199–200
  prefrontal cortex, 119
  rejuvenating with GHK, 229
  reticular activating system and
    trigeminal nerve, 178–79
  thallium damage, 131
brain-derived neurotrophic factor
  (BDNF), 117, 125–26
brain health and processes, 103–28
  overview, 103–4
  hacking the language processing area
    with a laser (not recommended),
    110–11
  hippocampal atrophy, 25–26, 107
  microglial cells, 17, 111, 115, 194
  and monounsaturated fats, 52–53
  neurofeedback, 105–9, 128
  red/infrared laser therapy for, 109–12
  and sleep, 64, 69–70
  See also cognitive dysfunction;
    neurotransmitters
brain herbs, supplements, and drugs
  overview, 116–17
  Brain Octane Oil, 57–58, 115, 154, 201
  coenzyme Q10/idebenone, 123–24
  curcumin, 126–27
  He Shou Wu, 127
  lion's mane mushroom, 125–26
  L-theanine, 125
  microdose deprenyl/selegiline, 122–23

  microdose nicotine, 120–21
  modafinil, 118–20
  piracetams, 117–18
  pyrroloquinoline quinone, 124–25
Bredesen, Dale, 104
Brigham and Women's Hospital,
  Boston, 11
Bulletproof Coffee, 57, 59, 60, 90, 115,
  139, 192–93, 201, 246
Bulletproof Diet, The (Asprey), 194
Bulletproof Diet and sleep needs, 67
Bulletproof Radio, xvi–xvii, xviii, 60
butter from grass-fed cows, 53, 55, 57
butylparaben, 168
butyric acid, 198

C360 Health, 261–62
cadmium, 34, 35, 130–33, 136, 141
calcification of tissue, 140–41
calcium, 140
calorie restriction, 154, 219
Campbell, T. Colin, 45, 46–47
Campbell, Thomas M., II, 45, 46–47
cancer
  contributing factors, 5
  and genetics, 18–19
  incidence of, 4, 18
  and middle-age hormone level
    reduction, 158–59
  modified citrus pectin vs., 138–39
  modified NK cells vs., 223–25
  and Naltrexone, 259
  from outdoor blue light, 79
  tumors overexpress substance P, 177
Candida fungus, 190–91, 192, 201
capsaicin for dental health, 181
carbohydrates
  effect of decreasing intake of, 16
  hormone depletion caused by, 166
  pairing with blood sugar reducing
    supplements or fiber, 113, 128
  "zero carbs for nineteen days"
    experiment, 69
carbon 60 and Carbon60 Plus, 260–62,
  263
cardiolipin, 237

cardiovascular disease, 12–13, 22, 131. *See also* heart disease; vascular system

carotenoid supplements, 82

cartilage growth with ozone therapy, 150

catalase, 241

cavemen and fire, xiii–xiv, 207, 208

cayenne pepper for dental health, 181

CDC (Centers for Disease Control and Prevention), 4

celiac disease, 42, 198

Cell Host & Microbe (journal), 198

cell membranes, 51, 52

cellular apoptosis, 9, 19, 25, 37, 38–39

Centers for Disease Control and Prevention (CDC), 4

CFL (compact fluorescent light) bulbs, 78, 81, 132

chelation therapy and chelating agents
overview, 34–35, 143
EDTA, 139–41
glutathione, 134–35

children and dogs, playing with, 165

China, xxi, 131

China Study, The (Campbell and Campbell), 45, 46–47

chlorella, 138, 143

cholesterol
and activated charcoal, 137
and low thyroid hormones, 172
testosterone made from, 166

chondroitin sulfate, 217

chromium, 130

chromium picolinate, 113

circadian clock disruptions, 79, 88, 108

circadian rhythm
overview, 84–85, 88
and blue light, 78–79
effect of food, light, and too little sleep, 64
and ketones, 60, 61
and red light, 89–90, 93
and vitamin D, 87

CLA (conjugated linoleic acid), 53

Clomid, 257

Clostridium difficile bacteria, 189

coconut oil, 57–58, 182

coenzyme Q10/idebenone, 123–24

cognitive dysfunction
overview, 104–5
Asprey's diagnosis of, 17–18
brain shrinkage, 25–26
and jaw misalignment, 180
laser therapy vs., 109–12
LED lights flickering vs., 111
from toxic mold exposure, 5
vascular issues and, 12–13
See also brain health and processes

cognitive enhancement, 30, 115, 120–21, 128. *See also* brain herbs, supplements, and drugs

collagen
overview, 233–36
and bakuchiol, 238
and cryotherapy, 236–37
and GHK, 229
and laser facials, 239–40
from light therapy, 94–95
and melanin, 240
and methylene blue, 238–39
and microneedling, 237
in pig's ears, 56–57
as protein powder, 48, 201, 234, 246
red light for stimulating production, 92–93
and retinol, 238
and vampire facials, 237
vitamin C for production of, 7

compact fluorescent light (CFL) bulbs, 78, 81, 132

conjugated linoleic acid (CLA), 53

Cook, Matthew, 223, 224

copper, 34, 130, 136

copper orotate supplements, 136

copper peptides (GHK-Cu), 229–30, 231, 249

cortisol, 70, 164, 171, 175

Crohn's disease, 252, 259

cryofacials, 236

cryotherapy, 236–37

curcumin/turmeric, 80, 126–27, 219

cyclical ketogenic diet, 114–15, 128

cypress essential oil, 225, 231
cytokines, 10, 13, 17, 175, 177
cytomaxes and cytogens, 249–50

DAO (diamine oxidase) enzyme, 192
death, human avoidance of, xiv, xv–xvii
death by a thousand cuts
  overview, 6, 20, 26, 101
  and brain function, 107–8
  defense strategy, 10, 96–97, 101
  inflammation factor, 108–9
  and mitochondria, 8–9, 108, 232
  sleep for recovery from, 67, 77
dehydroepiandrosterone (DHEA), 158,
  162–63, 173
delta (deep) sleep, 69–71, 74, 77–78
dementia. See cognitive dysfunction
de novo lipogenesis, 50
dental health hacks, 180–83. See also
  teeth
deprenyl (selegiline), 122–23
depression, 199
detoxing, 133–34, 142. See also heavy
  metal detox methods; waste buildup
  inside cells
DHEA (dehydroepiandrosterone), 158,
  162–63, 173
DHT (dihydrotestosterone), 168–69,
  242, 245, 246
diabetes, 4, 5, 13–17, 87
diabetes, type 2, 12–13, 33, 194. See
  also insulin resistance
diabetes, type 3, 14. See also
  Alzheimer's disease
Diabetes (journal), 32
diamine oxidase (DAO) enzyme, 192
diet
  anti-inflammatory, low in sugar, 183
  calorie restriction, 15, 154, 219
  cyclical ketogenic diet, 114–15, 128
  heart healthy, low-fat, low-cholesterol,
    166–67
  limiting eating window to 6–10 hours,
    49, 60–61, 62, 154, 155
  polyphenols and antioxidants, 246
  protein restriction, 47–49
  raw ominvore, 46

reducing the effect of carbohydrates,
  113, 128
relationship to health, 14–15
vegan, 45–47
Western diet, 166–67
See also fasting; food
digestive fiber, 138–39
dimmer switches for lights, 79, 81, 90
dismutase, 127
DMC (dimethoxychalcone), 30
DNA, 152, 200
DNA Company, 23
dopamine, 122–23, 127, 172

Ebola epidemic, Sierra Leone, Africa,
  148–49
eczema, 177
ED (erectile dysfunction), 121, 213,
  220–22
EDTA (ethylenediaminetetraacetic
  acid) chelation therapy, 139–41
EEG (electroencephalogram), 69, 105–7
electrical stimulation pill, 191
electron shuffling with NAD, 151–54
embryonic stem cells, 209–10
emotional stress, inflammation from, 34
endocrine system, IVL therapy for, 96
End of Alzheimer's, The (Bredesen), 104
endothelium, 10
energy and protein, 48
energy fats, 55–61
energy increases, 6. See also
  mitochondrial energy
Enig, Mary, 51–52
environment
  evergreen trees, 225, 231
  hormone-disrupting chemicals,
    167–68
  toxic mold exposure effects, 3, 5, 8,
    12–13, 20, 225
  See also death by a thousand cuts
enzymes, 122–23, 127, 192, 241
Eos and Tithonus, 265
epigenetic mechanisms, mitochondrial, 27
Epitalon, 38, 249
erectile dysfunction (ED), 121, 213,
  220–22

estrogen, 158–59, 168
"eternal life" quest, xiv
evergreen tree aromatic compounds,
    225, 231
EverlyWell, 32–33, 203
exercise
  addictive nature of, 121
  for boosting testosterone levels, 173
  choosing based on how well you
    slept, 83
  detoxing by sweating during, 141–42
  high-intensity interval training,
    141–42, 170–71
  and hormones, 170–71
  for increasing Klotho levels, 228
  Kegels, 222
  for preventing telomere shortening, 38
  and SARMS, 257–59
  weight lifting, 219
exosomes, 215–16
extracellular aggregates, 33–35
extracellular matrix stiffening, 31–33

fasting
  overview, 49
  for breaking down amyloids, 34
  and fasting-induced adipose factor, 197
  for heavy metal detox, 142
  for increasing NAD+ levels, 154
  intermittent fasting, 49, 60–61, 62,
    154, 155
  limiting eating window to 6–10 hours,
    49, 60–61, 62, 154, 155
  and stem cells, 219
fasting-induced adipose factor (FIAF),
    197
fats
  overview, 50–55
  energy fats, 55–61
  and healthy testosterone levels,
    166–67
  omega-6 and omega-3, 50, 53–55, 62
  replacing undesirable calories with
    saturated fats, 173
  saturated fats, 50–52
  types of, 51–52
fatty acids, 11–12, 48

fermented foods, 192, 193, 203
ferritin levels, 226
fertility, 170. See also bioidentical
    hormone replacement therapy;
    hormones
FIAF (fasting-induced adipose factor),
    197
fiber
  overview, 138
  modified citrus pectin, 138–39
  pairing carbohydrates with, 113, 128
  prebiotic fiber, 193–95, 198–99, 201,
    203
fibromyalgia, 259
field toxins, 42–43
fight-or-flight response on EEG, 106
fisetin, 30, 40
fish and heavy metals, 131
fluorescent lights, 86
flutathione, 134–36
food, 41–62
  overview, 41
  aging from, 69
  Big Food and glyphosate, 44
  energy fats, 55–61
  fats, 50–55
  fermented foods, 192, 193, 203
  grains, gluten, glucose and glyphosate,
    41–45, 62, 194–95, 201
  industrially-raised animals, 12, 43–44,
    47, 51, 201
  nightshade sensitivity, 181, 201
  pig's ears for collagen, 56–57
  See also diet; meat
food sensitivities, 34, 40, 198
Fountain of Youth, xvii
Four Killers, 3–23
  overview, 4, 21, 22
  and AGEs from fried or charred meat,
    36, 40
  and blue light exposure, 86–87
  common underlying issue, 8–10
  and mitochondria, 8–10
  prebiotic fiber vs., 194
  See also Alzheimer's disease; cancer;
    diabetes; heart disease
Fowkes, Steven, 116

free radicals (reactive oxygen species)
  overview, 7–8
  from blue light exposure, 86
  as byproduct of ATP production, 9
  and cancer, 18
  and mitochondrial mutations, 27–28,
    35–36
  ozone therapy vs., 151
  PQQ vs., 124
  released by MAO-B, 122
  and senescent cells, 28
fried, blackened, and charred meat,
  36–37, 40, 46, 54, 62
Fuller, Buckminster, 260–61
full-spectrum light, 86

GAINSWave shock-wave therapy,
  221–22, 231
galectin-3 molecule, 139
Gallagher, Dr., 144
Game Changers (Asprey), 41
gamma waves, 70
Gartenberg, Daniel, 77–78
gastroesophageal reflux disease
  (GERD), 199, 235
genetics
  functional genomics and risk, 22–23
  genotoxins, 43
  MITF gene, 241–42
  mitochondrial DNA, 26–28
  sequencing the human genome, 27, 102
genital stem cell therapy, 213, 214, 217,
  220–21
genotoxins, 43
GERD (gastroesophageal reflux
  disease), 199, 235
germs, 185–203
  microbes passed to you at birth,
    189–91
  microbiome, 185, 186, 187–88,
    189–91, 192, 199, 200
  See also bacteria; gut bacteria
GHK-Cu (copper peptides), 229–30,
  231, 249
ghrelin, 64
glucosamine and mitochondrial
  biogenesis, 15

glucose and blue light, 86
glucosepane, 32
glutathione
  overview, 135
  breaking down hydrogen peroxide, 241
  and cryotherapy, 236
  and hair growth, 245
  and heat shock proteins, 94
  for heavy metal detox, 134–35, 143
  vs. hydrogen peroxide buildup, 241
  liver support, 31
  and ozone therapy, 146
  and steep temperature drops, 236
gluten sensitivity, 42
glycated albumin, 194
glycation, 31
glycine, 43, 234–36
glymphatic system, 65
glyphosate, 42–44, 201
grain, 42–43, 62, 201
grass-fed animal products
  butter, 53, 55, 57
  collagen, 48, 62, 201, 246
  meats, 37, 167, 169–70, 173
  as protein source, 47–48
  switching to, 44
gray hair, reason for, 241
Greek mythology, xvii, 207, 227, 265
Greene, Robert, 1
green tea extract, 220
Grey, Aubrey de, 24
gut as an ecosystem, 186–87
gut bacteria, 185–203
  overview, 186–89
  and atherosclerosis, 11
  bacterial fuel and the gut lining,
    197–200
  bifidobacteria, 198
  and digestive fiber, 138–39
  and obesity vs. thinness, 197
  and parasites supplemented with
    helminth therapy, 187
  plasmid level exchange among, 11–12
  and resistant starch, 196–97, 203
  transforming veggies into fatty acids, 48
gut-brain axis, 199–200
gut lining, 197–98, 234–35

gut microbes, 198
gut microorganisms, tracking your, 200–202

hacks. *See* biohackers; biohacking
hair, 127, 168–69, 229, 241–45, 246–47
halogen light, 82
Harvard Medical School, 111–12
Hashimoto's thyroiditis, 5–6
Hayflick limit, 37
HCL (betaine hydrochloride), 235
head massages, 245, 247
healing and regeneration
  Herodotus on Fountain of Youth, xvii
  from inside out, 95, 232–33
  See also multiple methods for strengthening your body
healing peptides, 250–52
heartburn, 235
heart disease
  overview, 10–13
  and atherosclerosis, 10–11, 172
  incidence of, 4
  inflammation as cause of, 10–12
  and lead exposure, 131
  senile cardiac amyloidosis, 33
  and testosterone level, 159
heart health and activated charcoal, 137
heart rate variability (HRV), 71–73, 75
heavy metals, 34, 130–33
heavy metal detox methods, 133–42
  overview, 133–34
  activated charcoal, 136–38
  chlorella, 138
  citrus pectin, modified, and other fibers, 138–39
  EDTA chelation therapy, 139–41
  flutathione and other antioxidants, 134–36
  sweat it out, 141–42
helminth therapy, 187
hemophilia, transfusions for, 226
herbs
  ashitaba, 30
  astragalus, 39
  He Shou Wu, 127

stephania root, 127
  See also brain herbs, supplements, and drugs; supplements
Herodotus, xvii
He Shou Wu (polygonum multiflorum), 127
HGH (human growth hormone), 165–66, 170–71
HIIT (high-intensity interval training), 141–42, 170–71
hippocampal atrophy, 25–26, 107
histamines and histamine intolerance, 191–93, 203
hormetic stress, 48–49, 145–46
hormones, 156–73
  overview, 158
  alpha-MSH, 240
  cortisol, 70, 164, 171, 175
  driving the aging process, 158–59
  fasting-induced adipose factor, 197
  ghrelin, 64
  immune system support, 70
  leptin, 64, 171
  from L-tyrosine supplements, 171–72
  melatonin, 64, 79, 85–86
  modern living vs., 166–70
  thyroid-stimulating hormone, 243–44
  thyroxine, 243–44
  work-up, 5–6, 156, 160–61, 173
  See also bioidentical hormone replacement therapy
HRV (heart rate variability), 71–73, 75
human growth hormone (HGH), 165–66, 170–71
human nature
  avoiding that which hurts, 26
  death avoidance, xiv, xv–xvii
  fight-or-flight response, 106
  prioritizing survival over socialization, 13
hyaluronic acid, 217
hydrogen peroxide, 241

IBD (inflammatory bowel disease), 177, 199
IBS (irritable bowel syndrome), 199, 252
idebenone/coenzyme Q10, 123–24
IDE (insulin-degrading enzyme), 113, 115

immune system
  and autoimmune diseases, 188
  cytokines as immune response, 10,
    13, 17, 175, 177
  hormones that support, 70
  infant to child development of, 190
  IVL therapy for, 96
  and modified NK cells, 223–25, 231
  senescent cells of, 29
  and telomere shortening, 37
incontinence, 220–22
induced pluripotent stem cells (iPSCs),
  222, 223–24
industrially-raised animals, 12, 43–44,
  47, 51, 201
infants' microbiome, 189–91
infections
  dental caps over low-grade infections,
    144, 155
  ozone therapy vs., 147–48
  viral infections and gray hair, 241–42
inflammation
  from AGEs, 31–32, 36, 45, 46, 112, 139
  and amyloids, 34
  from blue light, 86
  and brain dysfunction, 108, 112–16
  from cross-linked proteins, 32
  and cytokines, 10, 13, 17, 175, 177
  from dental caps over low-grade
    infections, 144, 155
  effect of reducing, 16
  from galectin-3 molecule, 139
  and lipopolysaccharides, 198
  from mitochondria dysfunction, 10
  Naltrexone for, 260
  from omega-6 fats, 53
  from oxytocin decrease, 164
  and plaque in arteries, 10–11, 50
  from senescent cells, 28–29
  and snoring, 69
  and teeth, 174
  from wheat-stimulated zonulin
    releases, 42
inflammatory bowel disease (IBD), 177,
  199
infrared sauna, 94, 97, 141. See also
  red/infrared light

injections, self-administered, 251
Insilico Medicine, 186
insulin, 64, 112, 115, 256
insulin-degrading enzyme (IDE), 113,
  115
insulin resistance, 29, 112–13, 196
insulin sensitivity, 115, 196–97
intracellular aggregation, 35–36
intravenous laser therapy (IVL), 95–96
ipRGCs (intrinsically photosensitive
  retinal ganglion cells), 79
iPSCs (induced pluripotent stem cells),
  222, 223–24
iron, 130
irritable bowel syndrome (IBS), 199, 252
islet amyloid, 33
IVL (intravenous laser therapy), 95–96
IV NAD+ treatments, 153
IV sedation, 215

jaw alignment, 177–80, 184
Jennings, Dwight, 178
jet lag, 78
joint pain, prolozone for, 150, 155
Joovv red light therapy, 93
junk (blue) light, 78–82, 85–90, 97
junk buildup. See waste buildup inside
  cells

Kandel, Eric, 107
Keck School of Medicine, University of
  Southern California, 66–67
Kegel exercises, 222
ketones and ketosis
  overview, 50–51
  cyclical ketogenic diet for reducing
    risk of Alzheimer's, 114–15, 128
  inducing lipolysis, 142
  and intermittent fasting, 60–61, 62
  from MCT oil, 57–58, 183
  in moderation, 33–34, 61
  and NAD+ to NADH ratio, 154
KetoPrime supplement, 154
Khavinson, Vladimir, 249
kidney disease, 14, 43, 227
Killen, Amy, 217
Klinghardt, Dietrich, 169

Klotho, 227–28
Kuro-o, Makoto, 227

Lana, Dr. *See* Asprey, Lana
Lancet, The (journal), 130–31, 194
laser facials, 239–40
laser for muscle soreness, 91–92
laser therapy
  and central nervous system, 174–75
  for cognitive dysfunction, 109–12
  intravenous laser therapy, 95–96
  for muscle soreness, 91–92
lead, 35, 130–33, 136, 141
leaky gut syndrome, 127, 198–99,
  201, 234
lectins, 194
LEDs (light-emitting diodes), 78, 81
  86, 86
leptin, 64, 171
Life Cykel lion's mane extract, 126
light, 84–97
  overview, 84, 97
  beneficial sources, 91–96
  blue/junk light, 78–82, 85–90, 97
  and circadian rhythm, 84–85
  dimmer switches for light sources, 79,
    81, 90
  infrared sauna, 94, 97
  intravenous laser therapy, 95–96
  power of light therapy, 110
  shining light on the brain, 109–12
  sunlight, 82, 84, 87, 88, 97
  yellow/orange/amber light, 81, 90,
    94–95, 246
  See also red/infrared light
light filter apps, 82
light sleep, 70
light sources with full range of
  frequencies, 89–90
Lights Out (Wiley), 156
lion's mane mushroom, 125–26
lipopolysaccharides (LPSs), 42, 198
Lipton, Bruce, xviii
Lithgow, Gordon, 34–35
longevity and circadian rhythms, 60,
  64, 88
longevity as a choice, xv–xvi

LPSs (lipopolysaccharides), 42, 198
L-theanine, 125, 128
L-tyrosine, 171–72, 173
Lucy nicotine products, 121
lypolysis, 141–42
lysosomes, 35–37

macular degeneration, 79–80, 93
Madera, Marcella, 216–17
MAO-B (monoamine oxidase B),
  122–23, 127
massage, 165, 245, 247
MCP (modified citrus pectin), 138–39
MCT (medium-chain triglyceride) oil,
  57–59, 183, 201
meat
  amount to eat, 47–48
  fried or charred meat and AGEs,
    36–37, 40, 46, 54
  grass-fed animal products, 37, 167,
    169–70, 173
medical tests
  allergy tests, 7
  at-home lab tests, 203
  calcium score, 140–41
  cholesterol and triglyceride levels,
    54–55
  EEG, 69
  fasting blood sugar, 5
  for heavy metal toxicity, 134, 141, 143
  hormone levels, 5–6, 156, 160–61, 173
  insulin sensitivity score, 115
  SPECT brain imaging technique,
    17–18, 103, 180
melanin, 59, 240, 241
melatonin, 64, 79, 85–86
memory and memory consolidation, 70
menopause, 158–59
mercury, 35, 130–33, 138, 141
metabolic flexibility, 49, 61, 62, 128, 202
metabolism and vegan diet, 46
metabolites and fasting, 49
metabolites in gut bacteria, 190–91
metformin, 29, 40
methionine, 47
methylene blue, 238–39
methylparaben, 168

microbiome
  and antibiotics, 185
  diversity in, 186, 187–89, 192
  feeding good bacteria, starving bad
    bacteria, 201
  gut/brain connection, 199
  of infants, 189–91
  and stress, 199
  tracking, 200–202
  tracking your gut microorganisms,
    200–201
  Viome analysis results, 201–2
microglial cells in the brain, 17, 108,
  111, 115, 194
microneedling, 237
micronutrients, 167
middle sleep vs. deep sleep, 70
Miller, Philip, 5–6
Mitchell, Ian, 261–62
MITF gene, 241–42
MIT (Massachusetts Institute of
  Technology), 111
mitochondria
  overview, 7
  balancing free radicals and antioxidant
    production, 9
  blue light exposure vs., 86
  electron shuffling with NAD, 151–54
  and eye health, 80
  EZ water for, 92
  fight, flee, feed, reproduce messages
    from, xv, xviii
  and Four Killers, 8–10
  and gut bacteria, 11–12
  and insulin resistance, 112–13
  and methylene blue, 238–39
  mitochondrial mutations, 26–28,
    35–36
  and PQQ, 124
  sleeping better by strengthening, 65
  and steep temperature drops, 236–37
  training your brain to be resilient, 115
mitochondrial biogenesis, 15, 120
mitochondrial dysfunction
  overview, 20
  anaerobic metabolism as, 19
  bad sleep, 66

and cancer risk, 18, 19
  from heavy metal toxicity, 35
  from junk light, 82
  macular degeneration from, 80
mitochondrial energy
  overview, 7, 9
  from coenzyme Q10/idebenone,
    123–24
  heavy metals in our bodies vs., 130
  and mitochondrial mutations, 26–28,
    35–36
  and NAD, 151–52, 153
  and red/infrared light, 92, 110–11
  toxic mold vs., 7–8
  in youth vs. later years, 9–10
modafinil, 118–20
modified citrus pectin (MCP), 138–39
modified natural killer (NK) cells,
  223–25, 231
mold aflatoxin, 137
monoamine oxidase B (MAO-B),
  122–23, 127
monounsaturated fats, 52–53
mouth and gums, cleaning, 181–83, 184
mTOR (rapamycin), 29–30, 40
muscle growth with SARMS, 253,
  254–55
muscle soreness, laser for, 91–92

NAD (nicotinamide adenine
  dinucleotide), 151–54, 155
NAD+ to NADH ratio, 152–53, 154, 155
Naltrexone, 259–60, 263
Nature Medicine (journal), 226
Near Future Summit, San Diego, 111
nerve growth factor (NGF), 117, 124,
  125–26
nervous system
  overview, 71–72
  and myelin, 50
  and trigeminal nerve, 174–76,
    178–81, 184
neurofeedback, 105–9, 128
neurological dentistry
  overview, 174
  jaw alignment, 177–80
  and TMJ dysfunction, 179–80, 184

and trigeminal nerve, 174–76, 178–81, 184
neuroplasticity, 107
neurotransmitters
  dopamine, 122–23, 127, 172
  glycine, 234–36
NGF (nerve growth factor), 117, 124, 125–26
nickel, 130, 136
nicotinamide adenine dinucleotide (NAD), 151–54, 155
nicotine, 120–21
Nightline (ABC program), 119
nightshade sensitivity, 181, 201
Night Shift (Apple's warm screen colors), 81–82
nitric oxide, 92
NK (modified natural killer) cells, 223–25, 231
nootropics (smart drugs), 116–17, 159. *See also* brain herbs, supplements, and drugs

obesity, 5, 13–14, 64, 197
oil pulling for dental health, 181–83
Okinawa Institute of Science and Technology, 49
omega-6 and omega-3 fats, 50, 53–55, 62
opioid receptor antagonist, 259–60, 263
oral health from oil pulling, 181–83, 184
oral nicotine, 120–21
oregano, essential oil of, 183
Oura Ring sleep tracker, 71, 75–77, 83
ovulation tracking, Oura Ring for, 75
oxaloacetate, 154
oxidation, 7–8, 52, 53
oxidative stress, 128, 130, 145, 152
oxygen, medical grade, 147
oxytocin, 163–65
ozone layer, 144
ozone therapy, 144–55
  administration of, 145, 147, 149–51
  at-home use, 146–47
  benefits from, 145–46
  and Ebola epidemic, Sierra Leone, 148–49
  IV treatments, 155

risk to prescribing doctors, 148–49
sterilizing teeth with, 144
ten-pass treatment, 149–50

palmitate, 50
pancreas, 64
Panda, Satchin, 60, 64, 88
parabens, 168, 173
parasympathetic nervous system, 71–72
Parkinson's disease, 122
PCOS (polycystic ovarian syndrome), 169–70
peptides, 248–52, 263
perforin, 224
performance-enhancing substances, 253–54
perimenopause, 158–59
peripheral artery disease, 13–14
personal care products, 168, 173, 244, 245, 246
PET (positron emission tomography) scans, 126
PGC-1 alpha (peroxisome proliferator-activated receptor gamma coactivator 1-alpha), 120
pharmaceutical companies and not patentable drugs, 38, 261
phthalates, 168, 173, 246
physiology and stress, 37–38, 72–73, 175, 176–77
pig's ears for collagen, 56–57
pioglitazone (Actos), 219
piracetams, 117–18
Plante, Jim, 228
plasmid level exchange, 11–12
platelet-rich plasma (PRP), 237
pluripotent stem cells, 210
polycystic ovarian syndrome (PCOS), 169–70
polygonum multiflorum (He Shou Wu), 127
polyphasic sleep (Uberman Sleep Schedule), 66
polyphenols, 30, 40, 59, 197, 201, 203
population growth rate, xx–xxi
postural orthostatic tachycardia syndrome (POTS), 12–13

POTS (postural orthostatic tachycardia syndrome), 12–13
PPAR agonists, 219
PPL (piperlongumine), 30–31, 40
PQQ (pyrroloquinoline quinone), 124–25
prebiotic fiber, 193–95, 198–99, 201, 203
prediabetes, 5, 8, 13–14, 15
prefrontal cortex, 119, 180
pregnenolone, 158
prehormones, 158
premenstrual syndrome, 78
prescription medications
  Actos (pioglitazone), 219
  acyclovir, 242
  birth control pills, 168, 169
  as cause of sexual dysfunction, 221, 231
  Clomid, 257
  Epitalon, 38
  immune-suppressing drugs, 188
  metformin, 29, 40
  Naltrexone, 259–60, 263
  one-at-a-time method for trying, 117, 118
  oxytocin lozenges and nasal spray, 165
  racetams, 116–18
  rapamycin, 29–30, 40
  steroids, 162, 248, 253–54
  See also brain herbs, supplements, and drugs; SARMs
probiotics
  histamines and histamine intolerance, 191–93
  LKM512 from Japan, 188
  prebiotic fiber, 193–95, 198–99, 201, 203
Prometheus, 207, 208
protein
  inflammation from cross-linked proteins, 32
  Klotho, 227–28
  as poor fuel source for humans, 48
  sirtuins, 152
  type and amount to eat, 47–48, 62
  See also amyloid proteins; collagen
protein restriction diet, 47–49

PRP (platelet-rich plasma), 237
psoriasis, 177

racetams, 116–18
rapamycin (mTOR), 29–30, 40
rapid eye movement (REM) sleep, 69–71
reactive oxygen species (ROSs). See free radicals
red/infrared light
  overview, 81, 91, 97
  benefits prior to and after sunlight exposure, 87–88, 246
  as counter to blue light damage, 89
  infrared sauna, 94, 97
  laser therapy for human brain, 109–12
  most effective wavelengths, 92–93
  therapy choices, 91–93
relaxation techniques and heart rate variability, 72
REM (rapid eye movement) sleep, 69–71
resistant starch, 196–97, 203
resveratrol, 220
retinol, 238
RNA molecules, 200
ROI (return on investment)
  for longevity-based choices, xvi, xix–xx, xxi–xxii
  for sleep trackers, 68
  smoking or vaping, 121
ROSs (reactive oxygen species). See free radicals
Roundup, glyphosate in, 42–44
Rowen, Robert, 148–49
Russian electrical stimulating pill, 191
Russian scientists' discovery of IVL, 95–96
Russian Stick Bodywork, 218

SAD (seasonal affective disorder), 78
sarcopenia, 25
SARMs (selective androgen receptor modulators)
  overview, 252–55
  GW501516 (Cardarine), 257–58
  LGD-4033 (Ligandrol or Anabolicum), 256–57

MK-2866 (Ostarine), 255–56
SR9009 (Stenabolic), 258
saturated fat, 50–52, 58
saunas, 94
scoliosis, 179
seasonal affective disorder (SAD), 78
selective androgen receptor modulators.
    See SARMs
selegiline (deprenyl), 122–23
senescent cells (death-resistant cells)
    overview, 28–31
    boron for reducing inflammation from,
        219, 231
    increasing number of with age, 250
    and methylene blue, 239
    modified NK cells and perforin vs.,
        224–25
    and NAD+ levels, 153
    ozone therapy vs., 145–46
senile cardiac amyloidosis, 33
SENS (Strategies for Engineered
    Negligible Senescence) Research
    Foundation, 24
Seven Pillars of Aging, 24–40
    overview, 24–25, 39–40
    cellular straitjackets, 31–33
    extracellular aggregates, 33–35
    junk buildup inside cells, 35–37
    mitochondrial mutations, 26–28,
        35–36
    shrinking tissues, 25–26
    telomere shortening, 37–39
    zombie cells, 28–31
Sex, Lies, and Menopause (Wiley), 156
sex hormone binding globulin (SHBG),
    168–69
sexual activity
    as anti-aging strategy, 162
    hormones associated with, 167–69
    improving sexual function, 213, 214,
        217, 220–22
    and testosterone cream, 161–62
Shallenberger, Frank, 144, 145
SHBG (sex hormone binding globulin),
    168–69
shrinking tissues, 25–26, 35–36
Shriver, Maria, xvi–xvii

SIBO (small intestinal bacterial
    overgrowth), 195, 201
Sierra Leone, Africa, 148–49
Silicon Valley Health Institute (SVHI),
    6, 178
sirtuins, 152
skin care, 246. See also aging backward;
    collagen
sleep, 63–83
    overview, 63, 69
    after full-body stem cell makeover,
        217
    amount needed, 66–67
    benefits of, 64
    blue light vs., 78–82
    detrimental effects of too little,
        64–66, 68
    dial in your sleep state, 77–78
    evaluating the quality of your, 68
    and heart rate variability, 71–73
    jaw alignment for improving quality
        of, 178
    longevity and circadian rhythms, 60,
        64, 88
    quality over quantity, 77–82
    rem/slow-wave sleep, 69–71
    and stem cells, 220
sleep apnea, 65–66, 178
sleep tracking devices
    overview, 73–74
    Asprey's use of early models, 68
    Basis wristband, 68
    Oura Ring sleep tracker output, 71,
        75–77
    Sleep Cycle app, 74–75
small intestinal bacterial overgrowth
    (SIBO), 195, 201
Smart Drug News (newsletter), 116
smart drugs (nootropics), 116–17, 159.
    See also brain herbs, supplements,
    and drugs
snoring, 65–66, 69, 74, 178
Somatic Training Network, 218
Sonic Sleep Coach app, 77–78, 83
SPECT brain imaging technique,
    17–18, 103, 180
spermidine, 188

stem cell exhaustion, 211
stem cells, 209–31
  overview, 209–11
  exosomes from, 215, 216
  stimulating production and availability,
    218–20
  umbilical and amniotic fluid stem
    cells, 215–16
  US regulations, 216
stem cell therapy
  overview, 231
  cancer-fighting stem cells, 222–25
  cells injected into predetermined
    areas, 213–14
  full-body makeover, 214–18
  harvesting the stem cells, 211–12
  results, 214, 217–18
  for reversing hippocampal atrophy, 26
stephania root, 127
steroids, 162, 248, 253–54
stomach acid and heartburn, 235
storage toxins, 42–43
stress
  aging from out-of-control stress,
    106–7
  determining your physiological level
    of, 72–73
  and hair loss, 243–44, 246
  and heart rate, 72
  and heart rate variability, 72
  hormetic stress, 48–49
  Klotho levels lowered by, 228
  and microbiome, 199–200
  and NK function, 225
  physiology and, 37–38, 72–73, 175,
    176–77
  telomere shortening from, 37–38, 39
stress management, 40
substance P, 176–77, 211
sugar
  cutting back or out, 203, 219
  and microbiome health, 201
  oxidative stress from, 128
  See also blood sugar
sugar cravings, 44–45
sun damage, 88

sunlight, 82, 84, 87, 88, 97
Superman, 84, 129
superoxide dismutase, 122–23, 236
supplements
  overview, 163
  AEP, 29
  alpha-lipoic acid, 135
  betaine hydrochloride, 235
  boron, 219, 231
  carotenoid, 82
  chlorella, 138
  chromium picolinate, 113
  collagen, 234
  copper orotate, 136
  DHEA, 158, 162–63, 173
  fat soluble supplements, 50
  fisetin, 30, 40
  L-tyrosine, 171–72
  oxaloacetate, 154
  resveratrol, 220
  TA-65, 39
  Tru Niagen, 153
  turmeric/curcumin, 80, 126–27, 219
  vanadyl sulfate, 113
  vitamin A, 167
  vitamin C, 7, 135–36, 143, 220, 236
  vitamin D, 34, 40, 87, 140, 167
  vitamin D_3, 167, 173, 220, 228
  vitamin K_2, 140, 167, 173
  zinc, 136, 173
  See also brain herbs, supplements, and
    drugs; glutathione; herbs
suppositories, EDTA as, 140, 143
sweating for heavy metal detox,
    141–42
Sykes, Dan, 218
sympathetic nervous system, 71–72, 106
synthetic hormones, 160

TA-65 (cycloastragenol), 39
TB500 healing peptide, 250–52
teas, 57, 125, 220
teeth, 174–84
  alignment of, 175–76, 177–80
  bite guards, 178–79, 180–81, 184
  correct bite for, 179

dental caps over low-grade infections, 144, 155
keeping calcium in, 140
mercury in silver fillings, 131
overbite, 180
See also neurological dentistry
telomeres, 37–39, 152, 171
temporomandibular joint (TMJ) dysfunction, 179–80, 184
ten-pass ozone treatment, 149–50
TENS (transcutaneous electrical nerve stimulation), 184
testosterone
overview, 6, 159
bioidentical hormone replacement therapy, 161–62, 257
and exercise, 170–71
and Klotho level, 228
and SARMS, 256–57
thallium, 130, 131–32
Thomas, Dylan, ix
three Fs (fear, feed, f___), xv
thyroid function, 46, 171–72, 243–44, 247
thyroid-stimulating hormone (TSH), 243–44
thyroxine (T4), 243–44
TIMPs (tissue inhibitors of metalloproteinases), 222–23
tissue inhibitors of metalloproteinases (TIMPs), 222–23
Tithonus and Eos, 265
TMAO (trimethylamine N-oxide), 11, 185
TMJ (temporomandibular joint) dysfunction, 179–80, 184
TMS (transcranial magnetic stimulation), 77
tools for controlling our biology, xviii
torticollis, 179
toxic mold exposure effects, 3, 5, 8, 12–13, 20, 225
transcranial magnetic stimulation (TMS), 77
transcutaneous electrical nerve stimulation (TENS), 184

trans fats, 53–55
trigeminal nerve, 174–76, 178–81, 184
trimethylamine N-oxide (TMAO), 11, 185
TrueDark, 81
TrueLight amber light device, 95
Tru Niagen (nicotinamide riboside), 153
TSH (thyroid-stimulating hormone), 243–44
turmeric/curcumin, 80, 126–27, 219

Uberman Sleep Schedule, 66
ultraviolet A (UVA) and ultraviolet B (UVB) light, 87, 90
University of Alabama at Birmingham, 241–42
University of California, Los Angeles, 126
University of Colorado, Boulder, 11
University of Connecticut, Storrs, 11
University of Pennsylvania, 72
University of Sydney in Australia, 89
Upgrade Labs, Beverly Hills, 93, 153, 212
uranium, 130

vanadyl sulfate, 113
vanilla for dental health, 181
vascular system
and EDTA chelation, 140–41
IVL therapy for, 96
and nitric oxide, 92
stiffening from excess TMAO, 11, 185
vegan diet, 45–47
vegan trap, 46
very small embryonic-like cells (VSELs), 222–23
Viome, 32–33, 200, 201–2, 203
viral infections and gray hair, 241–42
virgin (young) blood
overview, 225–26
copper peptides in, 229–30, 231
visceral fat
and diabetes, 14
and eating fats, 54
visual acuity improvement, 223

vitamin A, 167
vitamin C, 7, 135–36, 143, 220, 236
vitamin D, 34, 40, 87, 140, 167
vitamin D$_3$, 167, 173, 220, 228
vitamin K$_2$, 140, 167, 173
VSELs (very small embryonic-like
    cells), 222–23

Warburg, Otto, 19
waste buildup inside cells, 35–37, 127.
    See also heavy metal detox methods
weightlifting magazine on sugar and
    carbs, 16
Weizmann Institute, Israel, 185–86
Western diet, 166–67
Wharton's jelly, 217
wheat protein, inflammation from, 42
WHO (World Health Organization), 43
Wiley, T. S., 156–57
Wiley Protocol, 157, 159
wisdom, sharing with others, xx

Women's Health Initiative, 159–60
World Health Organization (WHO), 43

XCT Oil, 182–83
xenobiotics, 141
XPRIZE Foundation, xxi
yak butter tea, 57

Yamasa Tokei Keiki pedometer, 73
yellow/orange/amber light, 81, 90,
    94–95, 246

Zak, Paul "Dr. Love," 164
"zero carbs for nineteen days"
    experiment, 69
zinc, 130, 136, 173
zinc orotate, 143
zombie cells, 28–31. *See also* senescent
    cells
ZONA Plus, 22
zonulin protein, 42